Tissue Science Technology

Dr. Neeraj Tandan

Tissue Science Technology

ISBN 978-93-5111-256-3

Published in 2014 in India by

RANDOM PUBLICATIONS

4376-A/4B, Gali Murari Lal, Ansari Road
New Delhi-110 002
Phone : +91-11-43580356, +91-11-23289044
e-mail: randomexports@gmail.com, sales@randompublications.com,
info@randompublications.com

Type Setting by : Keystoneprintads, Delhi-110051
Printed at Thomson Press (India) Ltd

Preface

The Bioengineering of tissues and organs, sometimes called tissue engineering and at other times regenerative medicine, is emerging as a science, as a technology, and as an industry. Tissue is a cellular organizational level intermediate between cells and a complete organism. Hence, a tissue is an ensemble of cells, not necessarily identical, but from the same origin, that together carry out a specific function.

An important component in the early development of tissue engineering was the parallel development of artificial biomaterials. In the mid-1960s, artificial skin for burn victims was being pursued as a symptomatic therapy, and later, synthetic fibers were being tried as artificial skin grafts for burn treatment. In the early 1970s, there were concerted efforts to treat artificial surfaces to be used in implants in ways that would enable them to avoid causing blood coagulation, by applying special heparin complex coatings, for example. Other efforts focused on the toxicology profiles and biocompatibility of a variety of organic polymers considered for implants or tissue engineering, and the development of novel gels as the basis of artificial skin. In the late 1970s, researchers experimented with collagen-based artificial skin for use in oral mucosa injuries.

Tissue culture and plant regeneration are an integral part of the most plant transformation strategy, and can often prove to be the most challenging aspect of a plant transformation protocol. The key to success in integrating plant tissue culture into plant transformation strategies is the realization that a quick (to avoid too many deleterious effects of somaclonal variation) and efficient regeneration system must be developed. However, this system must also allow high transformation efficiencies from whichever transformation technique is adopted. Not all regeneration protocols are compatible with all transformation techniques.

This technique is effective because almost all the plant cells are totipotent. In each cell possesses the genetic information and cellular machinery necessary to generate the whole organism. Since, this technique can be used to produce a higher number of plants that are genetically similar to a parent plant as well as to another. Two concepts, plasticity and totipotency, are the central processes to understand the regeneration and plant cell culture. Plants, due to its longer life span and sessile nature, have developed a greater ability to overcome the extreme conditions. Tissue engineering is a newly emerging biomedical technology and methodology to assist and accelerate the regeneration and repairing of defective and damaged tissues based on the natural healing potentials of patients themselves.

For the new therapeutic strategy, it is indispensable to provide cells with a local environment that enhances and regulates their proliferation and differentiation of cell-based tissue regeneration. Biomaterial technology plays an important role in the creation of this cell environment. This book will be useful to biotechnologists, biologists, agriculture scientists, researchers, teachers and students.

I thank all members of my team who have helped in the preparation of the book. My special thanks go to "Random Publications" who have published the book.

– Dr. Neeraj Tandan

Contents

1

Tissue Culture Science and Laser Applications Technology

To study the effects of conventional laser application on the retinal pigment epithelium (RPE) in a perfusion tissue culture model of porcine retinal pigment epithelium without overlying neurosensory retina. RPE with underlying choroid was prepared from enucleated porcine eyes and fixed in a holding ring. Specimens were then placed in two-compartment tissue culture containers and were cultured during continuous perfusion with culture medium at both sides of the entire specimen, the upper RPE and the lower choroid (12 specimens out of 6 eyes). Cultures were kept for 1, 3, 7 and 14 days and were examined histologically.

Laser treatment was performed on each tissue ring by application of 3 × 3 laser burns one day after culture began (argon ion laser, wavelength: 514 nm, pulse duration: 100 ms; spot size: 200 μm) using different energy levels (400–1,000 mW); (16 specimens out of 8 eyes).

During laser treatment a marked lightening of the RPE with centrifugal spreading was observed. Using higher levels of energy, a contraction of the RPE towards the centre of the laser spot was noticed. One day after laser photocoagulation histology revealed destruction of RPE; within 3–7 days of culture, migration and proliferation of neighboring cells was observed in several lesions. After 7 days the initial defect of the irradiated area was covered with dome shaped RPE cells and after 14 days multilayered RPE cells were showing ongoing proliferation. However, there were also cases without proliferation after laser treatment.

The non-treated, continuously perfused RPE showed regular appearance in histological sections: during the first 7 days of culture, light microscopy revealed a normal matrix with a well-differentiated RPE monolayer. Subsequently proliferation even without treatment was observed and after 14 days the RPE became multilayered. It was possible to study the early healing response to the effect of laser treatment using the permanently perfused tissue culture system. A marked proliferation and repair of the laser defect could be observed in several but not all lesions. After 14 days even without laser treatment a proliferative multilayered RPE was present. Although this limits the use of the system for longer than 7 days, it seems to be useful for investigation of RPE-related disorders. The retinal pigment epithelium (RPE) is a complex cell layer involved in many eye diseases such as age-related macular disease (AMD), diabetic maculopathy (DMP), proliferative diabetic retinopathy (DRP), or central serous retinopathy (CSR). In these diseases laser photocoagulation is known to have a therapeutic effect and is therefore widely used.

Despite this widespread clinical use, the exact mechanism of laser photocoagulation is not completely understood. It is known that the therapeutic effect derives not from the laser burn itself, but from the subsequent biological reaction. Laser application using 514 nm is absorbed up to 50 per cent by RPE cells resulting in damages with subsequent healing and therapeutic effects, widely used clinically today. The ability to maintain RPE cells in culture has important advantages for the understanding of the pathobiology of several ocular diseases. Especially in organ cultures, the interaction of different tissue cells can be studied.

The main difficulty encountered when culturing RPE is to retain an intact monolayer over a long period of time, which would be helpful for elucidating complete pathomechanisms of several diseases. However, most previous studies are of limited value because the RPE cells were maintained on artificial substratum (e.g. polystyrene culture disks) although it has been well established that the substratum has pronounced effects on cell behaviour. To avoid these limitations Del Priore and coworkers first developed an organ culture system to maintain monolayers of RPE still attached to Bruch's membrane which provided the opportunity to study the response of the RPE to external stimuli such as photocoagulation. Using this model it was demonstrated that laser photocoagulation led to an acute disruption of individual RPE cells and a separation of damaged RPE cells from Bruch's membrane.

Treated areas were covered with irregular mounds of RPE cells within 7 days, which was supposed to mimic the response of RPE following laser photocoagulation in vivo. For our approach to study laser tissue interaction a new model for cultivating RPE cells was employed. It was developed to cultivate tissue in an organotypical environment of a two-compartment system under permanent perfusion with medium. Good results for preservation of adult full-thickness retinal tissue using this model were presented recently. In a second step we used this model to maintain RPE as an organ culture over a longer period of time and to characterize the vitality of this system by irradiating the RPE monolayer with a conventional argon laser and performing light microscopy afterwards.

TISSUE PREPARATION

Fresh enucleated porcine eyes were prepared with the use of an operation-microscope under sterile conditions according to the method developed by Kobuch et al. using the MinuCell-perfusion system. Thus 14 eyes were cut parallel to the limbus at a distance of 4 mm behind the iris-lens-diaphragm and the whole anterior cap was removed and most of the adherent vitreous. Next the posterior eye cup was separated into two equal parts by dissecting along the median raphe and the optic disc with a sharp knife. Each part was shortened at the edges to become a flat specimen of sclera, choroid, RPE and retina with partly adherent vitreous. The specimen was fixed to a flat surface with pins.

Neurosensory retina was peeled off from the RPE with tweezers. Small scissors were used to separate the choroid with the overlying structures from the sclera leaving a layer of choroid and RPE. In order to cultivate this matrix we used a ring holder system, placed into a perfusion container which has been previously described in detail. Briefly, the cell carrier consisted of a black holding ring, in which the cell layer was placed.

The internal diameter was 9 mm and the external diameter 13 mm. To fix the support, a white span ring was pressed into the holding ring. Thus the white ring was moved under the choroid and the black ring was put onto the RPE in order to fix the specimen into this ring system thus having choroid at the lower and RPE at the upper side. This was then placed into the container which was designed for one to six ring systems.

By placing the ring system into one of the compartments of the container, two compartments were created separated by the tissue and the ring holder system. Both compartments had two openings to be perfused with medium.

The use of the gradient container allows permanent perfusion of different media at the upper (retina) and lower (choroid) side of the sheets. In our study the same medium was used at both sides.

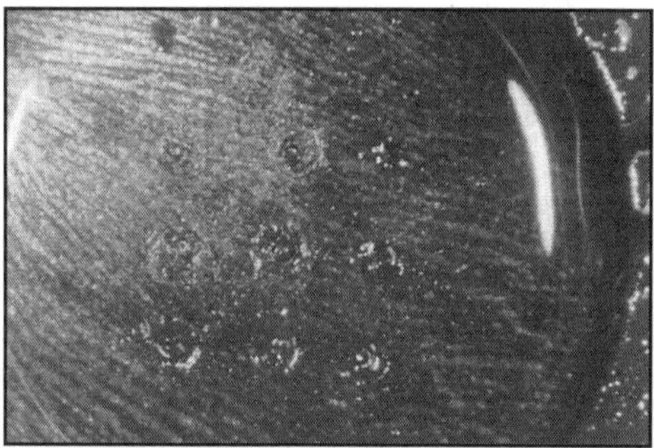

Fig. Tissue Holder Ring System with Coagulated RPE-Monolayer on the Top. Nine Laser Lesions can be Noticed

Over silicone tubes and luer fittings the container was connected to specific screw caps and media bottles allowing the permanent perfusion of the tissue. The system ran with a peristaltic pump with a speed of 1 ml/h on a warming table (40°C) out of the atmosphere of a CO_2-incubator. DMEM (Life Technologies) with 25 mmol HEPES, 15 per cent porcine serum and 1 per cent penicillin-streptomycin out of 10.000 IU was used.

The specimens were cultivated under these conditions for up to 1, 3, 7 and 14 days. Additionally one specimen was cultured up to 28 days. All were examined histologically by light microscopy.

Table. Number of Examined Specimens During the Study Differentiated by Perfused RPE without Laser Treatment and Perfused RPE with Laser Treatment: Every Specimen Treated by Laser Photocoagulation Covered Nine Spots

	1 day	3 days	7 days	14 days
Perfused RPE without laser	3	3	3	3
Perfused RPE with laser	4	4	4	4

CULTURE MEDIA AND CONTAINERS

The composition of culture media used for shoot proliferation and rooting has a tremendous influence on production costs. he type of culture vessel influences the efficiency of transfer during subculture and production of propagules per unit area.Proper choice of media and containers can reduce the cost of micropropagation. The replacement of expensive imported vessels with reusable glass jars and lids, alternatives to gelling agents, use of household sucrose, and some medium components can reduce costs of production Bulk making of media and storage as deep frozen stocks also reduces labour costs.

In vitro growth of plants is largely determined by the composition of the culture medium. The main components of most plant tissue culture media are mineral salts and sugar as carbon source and water. Other components may include organic supplements, growth regulators, a gelling agent. Although, the amounts of the various ingredients in the medium vary for different stages of culture and plant species, the basic MS and LS are the most widely used media.

During the past decades, many types of media have been developed for in vitro plant culture. Media compositions have been formulated for the specific plants and tissues. Some tissues respond much better on solid media while others on liquid media. In general, the choice of medium is dictated by the purpose and the plant species or variety to be cultured. The ratio of auxins to cytokinins in the culture medium is important since their combination determines the morphogenic response for root and shoot formation.Plant extracts such as coconut milk, banana extract, and tomato juice can be very effective in providing undefined mixture of organic nutrients and growth factors.A variety of other media components have also been used for specific purposes. The osmolarity of the culture medium, agitation, and aeration of suspension cultures have an important influence on plant cell division. The medium can be solid, semi-solid or liquid, depending on the presence or absence of gelling agents.

LOW COST MEDIA OPTIONS

Media chemicals cost less than 15 per cent of micro-plant production. In some cases the cost may be as low as 5 per cent. Of the medium components, the gelling agents such as agar contribute 70 per cent of the costs. Other ingredients in the media - salts, sugar and growth regulators - have minimal influence on production cost and are reasonably cheap. However, low cost options are available to replace expensive gelling agents, sugars and reduce the cost of water. For example, the water from distillation apparatus or passing through Millipore filters can be replaced with ordinary tap water in some cases.

GELLING AGENTS

The growth of cultures and production of shoots or roots is strongly influenced by the physical consistency of the culture medium. Gelling agents are usually added to the culture medium to increase its viscosity as a result of which plant tissues and organs remain above the surface of the nutrient medium. Many gelling agents are used for plant culture media, e.g., agar, 'Agarose', and 'Gellan gum', and are marketed under trade names such as 'Phytagel, Gelrite', and 'Gel-Gro' (ICA Biochemicals). Agar is the most commonly used gelling agent for preparation of solid and semi-solid media.

It contributes to the matrix potential, the humidity and affects the availability of water and dissolved substances in the culture containers. Various brands and grades of agar are available commercially, which differ in the amounts of impurities, and gelling capacity. Agar brands vary widely in price, performance and composition. It is the actual use and experience, which ultimately determines the choice of agar brand in a specific system and for a plant species. It is usually unnecessary to use high purity agar for large-scale micropropagation; cheaper brands of agar have been successfully used for industrial scale micropropagation.

The lowest concentration of agar, which can be used, depends on its purity and brand. Agar is usually used at 0.6-0.8 per cent (w/v). It is advisable to prepare sample media in small quantities using various concentrations of agar, e.g., 0.7, 0.75, 0.8, 0.85 and 0.9 per cent. The appropriate concentration should then be used for large-scale production purposes. In addition to cost saving, there are a number of other advantages in using low concentrations of gelling agents. A semi-solid medium ensures adequate contact between the plant tissue and the medium. It is beneficial to growth as it allows better diffusion of medium constituents, and is easily removed from plantlets before their transfer to in vivo conditions. For these reasons, a semi-solid medium is often preferred over solid medium.

Alternatives to Agar

Cheaper alternatives to agar include various types of starches and plant gums. The National Research Development Corporation, India has listed low cost agar alternatives, which are worth evaluating for routine use in commercial micropropagation. Gelrite can be replaced with starch-Gelrite mixture. The use of liquid media eliminates the need of agar. Other options include white flour, laundry starch, semolina, potato starch, rice powder and sago.

For micropropagation of ginger and turmeric, the combination of certain gelling agents gave growth as good as on agar-based media. The use of laundry starch, potato starch and semolina in a ratio of 2:1:1 reduced the cost of gelling agent by 70-82 per cent. However, the addition of such gelling agents to the medium also has some disadvantages. Some gelling agents contain inhibitory substances that hinder morphogenesis, and reduce the growth rate of cultures. Moreover, toxic exudates from the cultured explants may take a longer time to diffuse. Media solidified with gelling agents increase the time to clean the culture containers.

Corn-starch (CS) as a gelling agent has been used along with low concentration of 'Gelrite' (0.5 g 'Gelrite' + 50.0 g CS/l) for the propagation of fruit trees, such as apple, pear and raspberry, banana, and sugarcane, ginger and turmeric. The shoot proliferation was better on corn starch-medium than on agar. The cost of CS was $1.8/kg compared with $200/kg of agar. However, it became difficult to detect the contamination because the CS medium turned grayish-white.

Addition of 8.0 per cent tapioca starch to the MS medium was found to be a good substitute for 'Bacto-agar' for potato shoot-culture. Barley starch (60.0 g/l) has also been used for culturing potato-tuber discs, and for anther culture of barley. Sago (obtained from the stem pith of Metroxylon) at 13 per cent concentration was substituted for agar in MS medium for the multiplication of chrysanthemum. The number of shoots and leaves, and root length were significantly higher on sago than on agar. The cost of sago is $0.5/kg. 'Isubgol', a colloidal mucilaginous husk (chiefly composed of pentosans) derived from the seeds of Plantago ovata, has a good gelling activity, and has reasonable clarity in gelled form. 'Isubgol' at 3 per cent in MS medium has been used for the propagation of chrysanthemum. The cost of 'Isubgol' is about $4/kg.

Table: Effect of Agar Alternatives on Medium Solidification and Culture Growth

Agar alternative	*Effect on medium and culture*
Wheat flour (10%)	Sloppy medium, growth of cultures poor
Wheat flour (8%)	Proper solidification, growth of cultures below average
Laundry starch (6%)	Proper solidification, growth of cultures average
Semolina (5%)	Sloppy medium, growth of cultures average
Potato powder (7%)	Sloppy medium, growth of cultures average
Rice powder (11%)	Sloppy medium, growth of cultures poor
Sago (7%)	Proper solidification, growth of cultures normal
Laundry starch + Potato powder +	Solidification, growth of cultures as good as
Semolina (2:1:1)	On agar media

Use of Liquid Media and Physical Matrices

Suspension cultures without gelling agents are commonly used for culturing callus, cell clusters, buds and somatic embryos. Suspension systems allow greater contact between the explant and the medium. Moreover, agitation of such media reduces the diffusion gradient in the nutrient supply. The toxic metabolites exuding from the tissues are also dispersed effectively.

Table: Low cost Matrices

Option	*Use*
Cotton fiber	Callus maintenance and shoot organogenesis
Glass wool	Multiplication of chrysanthenum
Nylon cloth	Multiplication of chrysanthenum
Polystyrene foam	Multiplication of chrysanthenum
Glass beads	Multiplication of raspberry and white clover
Filter paper	Multiplication of chrysanthenum, potato

Strips (2.5x15 cm) of glass wool, nylon and filter paper have been used as supporting bridges for the propagation of chrysanthemum on MS medium. The shoot- and root-growth was almost identical on these matrices. Polystyrene foam blocks (2x2x1 cm) have been used as supporting matrix for the propagation of chrysanthemum.

The growth response was similar to that on glass wool, except that explants on foam produced significantly fewer roots but longer shoots than on glass wool. Glass bead based liquid-media were successfully used for culturing ginger and turmeric, and on per plant basis reduced the medium cost by 94 per cent With glass beads, the amount of medium required is only 15-18 ml per culture container.

By using 20ml medium per culture container, one liter media will give 50 culture containers, a substantial saving in medium cost. Ginger and turmeric plants multiplied on glass bead liquid-medium performed as good as on agar-based medium. A similar type of response was observed for vanilla.

Ficus cv.'mini lucii' showed higher multiplication rate although with a slight vitrification. Saintpaulia, Syngonium, Philodendron and Spathiphyllum also showed higher multiplication rates and better growth on glass bead liquid-medium. An additional advantage of liquid media withglass beads is that the medium can be replaced without moving plants.

Glass beads have been used for the propagation of raspberry and white clover resulting in 60 per cent saving in media cost. The glass beads allowed easy removal of plantlets from the medium. Glass beads have been also used for the maintenance of callus and shoot organogenesis. The beads can be reused after washing with acid. The liquid-media have some disadvantages. Delicate tissues get damaged during agitation. In some species, shoots submerged in liquid media become hyperhydric (water soaked), and unsuitable for micropropagation.

Conventional Carbon Sources and Low Cost Alternatives

Sucrose is the most commonly used carbon source in the micropropagation of plants. Sucrose adds significantly to the media cost. Household sugar and other sugar sources can be used to reduce the cost of the medium. Sugar sold in grocery stores is sufficiently pure for micropropagation. For culturing ginger and turmeric, all other carbohydrates except sugarcane juice, were suitable alternatives to laboratory grade sucrose. Use of common sugar reduces the cost of the medium between 78 to 87 per cent.

The cost of the local sugar was US$ 0.55/kg against the $40.0/kg for the imported sucrose. In Bangladesh, several laboratories have used locally available household sugar for culturing potato, banana, orchids, chrysanthemum, lentil, peanut, chickpea, medicinal plants, and fruit trees. In Pakistan, local sugar was found to be as good as the high grade laboratory sugar for the multiplication of banana. Maple syrup has been used for the multiplication (50 g/l) and rooting (34 g/l) of cherry root stocks from nodal segment and shoot tips.

Table: Effect of Alternatives to Analytical Grade Sucrose on Culture Growth

Alternative	*Effect*
Household sugar (3%)	Healthy cultures
Double refined sugar (3%)	Healthy cultures
Sugar crystals (3%)	Healthy cultures
Sugarcane juice (10% v/v)	Drying of leaf tips
Sucrose LR grade (3%)	Healthy cultures

Alternatives to other Ingredients

Several other ingredients can also be replaced by low cost options Commercial grade chemicals of lower purity than the analytical grades are quite suitable for commercial micropropagation unless deleterious effects are observed. A high degree of purity is justified only in the case of basic studies in tissue culture. In general commercial micropropagation, the quality will hardly be affected ordinarily by purity of these chemicals. Of all the chemicals, growth regulators (hormones) are the most expensive. However, they are needed in very small amounts in the medium, thus have a little effect on the medium cost. Sugar cane molasses can provide many of the nutrients,namely, sugar, vitamins and inorganic metal ions required for sugarcane callus induction and shoot formation.

At Dhaka University, low cost media have been developed for the multiplication of orchids based on macro-salts of any of the Vacin and Went, Phytamax and Knudson's C media, supplemented with 10-15 per cent (w/v) banana extract and 10-15 per cent (v/v) coconut water.

Use of above media eliminated the need of micro-salts and vitamins for the micropropagation of different orchid species. Several different orchids, namely, Vanda, Dendrobium, Aerides, Acampe and Spathoglottis can be multiplied on a medium containing only peptone, inositol, banana extract and coconut water. In many tropical orchid species, similar response was observed on media containing coconut water (15 per cent) and banana pulp (100 g/l), which effectively lowerd the cost.

Table: Low Cost Option for Sugar in Medium

Sugar type	*Use*
Refined white sugar (RWS)	Culture of zygotic embryos
Unrefined light brown sugar	Culture of zygotic embryos
Unrefined brown sugar,	Culture of zygotic embryos
Sugar maple syrup	Multiplication and rooting of cherry rootstock. Replace micro-nutrients and reduce macro-nutrients
Table Sugar	Multiplication of banana, potato, orchids, chrysanthemum; shoot regeneration and rooting of lentil, peanut, chickpea

Source of Water

Water is the main component of all plant tissue culture media. Usually in tissue culture research, distilled or doubled distilled and de-ionized water is used. Distilled water produced through electrical distillation is expensive. In some cases, alternative water sources can be used to lower the cost of the medium. If tap water is free from heavy metals and contaminants, it can be substituted for distilled water. Tap water has been used for in vitro propagation of banana and ginger, Zingiber officinale.

Table bottled water from the supermarket can also be used a low cost alternative. However, its mineral composition should be taken into account as it may affect pH and nutrient uptake. In rural areas, rainwater can be collected in clean glass jars and used for tissue culture. In Bangladesh, the change over of water distillation from electrical to gas operated unit reduced the cost from US$260 to $5/month for producing 50-60 liter water per day.

Low Cost Option for Media making

Media preparation is a time consuming process. Many companies sell ready-made media in liquid or powder form. The pros and cons of laboratory making verses buying readymade media should be considered. Although pre-made media save time, their relative cost is high. Pre-packaged media preparations are useful when medium quantities required are small. The errors in media preparation from pre-packed media are also less frequent. However, for large-scale use, it is much more economical to prepare media by mixing the basic ingredients. By making large batches, the time for media preparation and sterilization can be reduced substantially. However, if mistakes occur, the bigger is the batch, the greater is the loss. The volume of medium prepared in large sized-laboratories can range from 200 to 500 liters per day. The media should be made and used in small batches. Non-sterile media should not be stored at room temperature for more than 24 hr, especially when it contains high amounts of sucrose. These should be refrigerated until further use. Media are usually dispensed into containers either manually or with a peristaltic pump. Manually, one person can fill more than 200 containers an hour. In some large-scale laboratories, the distribution of sterile media is automated. To be economical, the automation must be able handle many hundred containers per hour.

Handling Stock Solutions

Usually the stock solutions are prepared in 10x to 1000x concentrations. The stock solutions consist of groups of chemicals, e.g. macronutrients, micronutrients, vitamins and plant growth hormones. The inorganic chemicals and vitamins solutions can be combined into a single, 10 X concentrated, stock solution. The stock solutions can then be frozen in plastic bags in volumes sufficient to make 1 to 5 liter media at a time. The cost of the medium prepared in the above way is much cheaper than the ready-made powdered media. This is being practiced for long time in many developing countries.

Cost of Medium

Depending on the plant species, the cost of culture medium varies with the ingredients such as sugars, gelling agents, and growth hormones used. Based on the current price of the various ingredients, the estimated cost of one-liter MS medium works out to be $0.18 for the solid and $0.08 for the liquid medium. If sucrose were not added, as in the case of autotrophic micropropagation, cost per liter would be $0.13 for the solid and only $0.03 for the liquid medium. The cost of medium per plant depends on the amount dispensed and the rate of plant multiplication per culture container.

Autotrophic Micropropagation

Plants with functional chloroplasts can grow in vitro on media without sugar, provided the micropropagation environment is modified to enable photosynthesis. The growth of plants on sugar-free medium, but with the carbon dioxide enriched environment was similar to sugar-containing media. Plants such as carnation, chrysanthemum, Cymbidium, Primula, potato and strawberry have been successfully grown using the above technique, termed as PTCS-'Photoautotrophic Tissue Culture System'. In the autotrophic system, plants are grown in large containers where the air content (oxygen, carbon dioxide, relative humidity, etc) and the composition of the culture medium is easily controlled. This system has several advantages that include reduction in production cost because of large culture vessels, simple culture media formulations, and lower incidence of culture contamination. Plants grown in such strongly aerated vessels require little or no hardening.

CULTURE CONTAINERS

A wide variety of culture containers are available on the market. The plant performance and cost of the containers should be used as the prime criterion in choosing the appropriate type. Depending upon the scale of production, different types of containers are deployed for culture initiation, maintenance of mother cultures and sub-culture for multiplication. For shipment of plants, disposable containers should be bought as and when required. Irrespective of the type, the containers used for maintaining in vitro plants should be transparent to facilitate illumination and easy inspection.

Types of Culture Containers

Glass test tubes have been universally used for culturing plant tissues as their narrow openings keep out contamination. Usually, only 1 to 2 explants are cultured in each test tube. As explant growth and multiplication progress, the cultures are transferred to larger containers. Test tubes should not be used for large-scale multiplication of plant material. Conical flasks, 150-250 ml capacity, are mostly used for the micropropagation of plants in liquid media, and are kept on shakers for agitation of the medium and suspension cultures.

However, conical flasks are more expensive than glass bottles, and their narrow mouths make manipulations of cultures difficult. Glass and plastic Petri dishes are also used to culture explants, which are then transferred to larger vessels for multiplication and elongation. Glass Petri dishes have to be sterilized before pouring medium that has to be sterilized separately. Glass Petri dishes are more expensive than the test tubes and conical flasks. Pre-sterilized disposable-plastic Petri dishes are much cheaper. In both cases, medium is poured under the laminar flow hood.

Glass bottles and baby-food jars with polypropylene caps are the most widely used containers and the most economic and low cost option. The wide mouth makes culture manipulation easy and approximately 15-20 explants can be inoculated in each bottle. Such containers are widely available and cost ca. US$0.09. Such containers can be washed, sterilized and reused repeatedly. In Cuba, glass containers are washed with a circular nylon brush mounted on an electric rotor, and then dried in the sun by inverting them on wooden trays. The containers are then double wrapped in newspapers and sterilized in an autoclave.

The medium is sterilized separately in flasks, and then dispensed into the containers under the laminar flow cabinets. Transparent plastic containers, such as 'Magenta' vessels, that withstand autoclaving and washing, are extensively used for micropropagation in many developed countries, but at US$1.5 to 2.0 are very expensive. Also, the repeated autoclaving of the plastic containers renders them cloudy, thus reducing the passage of light. Lately, containers are being made of polypropylene and polycarbonate and more recently polystyrene that can be autoclaved. Disposable, non-autoclavable food containers and sandwich boxes made from polystyrene have been also used for plant micropropagation. Use of disposable containers eliminates the cost of washing. PVC pots and jars, manufactured for the food-industry, have been used for in vitro plant culture.

The high temperature during their manufacture imparts relatively high degree of sterility, so that they can be used directly without sterilization. Hence, only the medium needs to be sterilized, thus eliminating the costs of container sterilization. Non-sterile disposable plastic type containers and bags can be bulk sterilized with gamma rays at industrial irradiation facilities. Such gamma-irradiated containers are being used for large-scale micropropagation. Being lightweight material, these can be shipped to distant places at a very low cost.

A new type of culture vessel called 'StarPac' TM has been manufactured. It is a disposable bag, and is semi-permeable to gas but requires different methods of handling in contrast to test tubes and other containers. Disposable plastic type containers called 'Watson Modules', have been used for production of micro- and mini-tubers of potato. The modules allow in vitro multiplication, plant hardening and soil growing in the same container, and permit batch handling of cultures during all stages of plant growth. Disposable containers have been also made from fluorocarbon plastic films, which can be autoclaved, and are transparent and impervious to water, but fairly permeable to gases.

Vessel Closures and Lids

The vessel closures influence growth of the in vitro plants. The culture containers have to be closed to keep out microorganisms but they should not be sealed so tightly that gaseous exchange is prevented. The different types of closures include non-absorbent cotton plugs, Polyurethane foam plugs, specially made plastic plugs, aluminum foil, stainless steel caps, Polypropylene caps with bends, PVC film, Polythene film and silicon rubber. Culture containers such as tubes and flasks are closed with non-absorbent cotton wool, sometimes wrapped in muslin gauze. But under large-scale production, preparing cotton plugs becomes cumbersome and time consuming.

Hence, these have been replaced with autoclavable screw caps made of stainless steel or polypropylene. Test tubes covered with translucent caps, which facilitate good gaseous exchange, have been used for culturing carnation. Polypropylene film, which is heat resistant, has been successfully used as closures. It is transparent, non-greasy, and can be used to seal large containers.

Sandwich wrapping films, such as 'Cling film' has also been used either on its own or as double wrap on top of the lids to prevent contamination after subculture. This is particularly suitable for mother culture, which may have to be maintained for 3 to 6 months or longer.

Gaseous exchange is important for the availability of oxygen and carbon dioxide, water vapour and elimination of ethylene gas build-up that is detrimental to plant growth. The shape of the container influences the growth rate of cultures by modifying gaseous diffusion. The best size and shape of the culture container varies with the plant species, and hence should be decided after prior experimentation.

It is important to ascertain the price, costs for washing, and the number of times the culture container can be reused. The selection of the containers should be done after making sure that they would not affect growth in vitro. Culture containers can be vented by modifying lids. to facilitate optimal growth. Depending on the plant species, microporous membranes of different sizes are available.

Low Cost Options for Containers

Plastic bags (approximately 10x15cm) have been used for large-scale micropropagation and are very cost effective. Disposable plastic bags eliminate the cost of washing and of lids. Plastic bags are sterile due to high temperature during manufacture. After pouring presterilized medium under the laminar flow, the top 2 to 5 cm of the bags is folded and several bags are held together with either large paper clips or plastic cloth-hanging pegs. After transfer of the explants and cuttings, the bags are closed with a heat-sealing machine or by knotting if the bags are 18-20 cm long. Plastic bags being lightweight can also be hung by thread and do not need elaborate shelving.

In Bagladesh and India, juice, jam, and jelly bottles have been used. The orchid producers in Thailand, Singapore, Indonesia, Malaysia, and India have been using old whisky bottles for orchid culture for many years. The production cost of tissue-cultured plants can be reduced by 50-90 per cent by using low cost media-ingredients and containers. A private company in India has been producing more than five million plants of various ornamentals (foliage and flowering), plantation, medicinal, farm and forestry plants using low cost media ingredients and containers.

LOW COST ENERGY OPTIONS

USE OF NATURAL LIGHT

Artificial lighting of cultures in the growth rooms is one of the most expensive and inefficient methods in tissue culture technology. The lights, chokes, fixtures, timer controls, equipment to handle high electrical load, and their operation and maintenance add to high costs. Moreover, artificial lighting generates heat that has to be dissipated by cooling and airconditioning further adding to the electrical load. Although special fluorescent tubes are used to compensate for the red and far-red part of natural daylight, artificial light quality does not match that of natural light under which the plants are ultimately grown. Also, the cool fluorescent lights used for illumination provide minimal energy required for photosynthesis. As a result, in vitro plants adapt to low-light intensity, and have a reduced growth rate.

Plants can adapt to a wide range of conditions by changing their metabolism and structures. They develop structural and anatomical features in leaves and stems, e.g. cuticle and wax on leaves, thickness of the leaf, fewer and closed stomata, and thickening of epidermal cells of the stems, that allow survival in the harsh environment. However, once they adapt to a set of conditions, re-adaptation to new conditions is rather slow or difficult. Plant tissues formed and adapted to low light conditions are usually fragile and may become vitrified, leading to poor survival under field conditions. This can be a major disadvantage in the plant hardening process, and later establishment in the field. Under artificial light of low intensity, plants have low reserves, and a poor root system.

On transfer to soil, the in vitro formed roots have to adjust to soil solutes of varying pH. The usual response of the in vitro formed roots is that they stop functioning in soil and new roots are formed, which take over the function of the original roots. If new roots do not emerge, the plant dies.

One method to circumvent these negative effects is to culture the in vitro plants under natural light, during their last phase in liquid medium, based on half- or quarter-strength MS salts without sugar and vitamins, under either aseptic or non-aseptic conditions. If roots or root initials are not formed, the medium can be either supplemented with auxins (IAA, IBA), or shoots dipped in a solution of rooting hormones. This procedure provides much stronger and healthier plants with a high survival rate.

Low Cost Lighting

Changing the method of illumination from artificial to natural light is a decisive low cost option in tissue culture. This not only reduces electricity and capital costs, but also improves the plant quality. Expensive artificial lights can be replaced in several ways. One option is to grow the in vitro cultures in diffused natural light under plastic or glass.

This works very well in temperate climates, but under tropical conditions, heat build-up has to be reduced by installing thermostat-controlled exhaust fans. It is also possible to redesign the existing laboratories in tropical and subtropical countries, and replace the artificial lights with 'Solatube' which can be installed in any roof, and redirects the daylight through reflective tubing. Such a tube currently costs ca. US $600, and once installed can illuminate 3-5m2 without incurring further running costs.

Yet another method is to incorporate Southwest facing glass windows in the growth rooms that allow indirect diffused natural daylight. By using either muslin or Venetian curtains made of bamboo or plastic, the light can be diffused to the desired intensity. The growth room is usually located on the upper storey of the house to allow sunlight from all directions. Natural light has been successfully used in this manner in "Bio-factories" in Cuba, based on the conversion of village houses into tissue culture laboratories.

In vitro cultures have been also successfully grown under natural light by using simple plastic bags (polypropylene bags) as culture containers. These bags are lightweight and can be hung in a greenhouse. They are transparent, so that incident illumination reaches the cultures with the least loss, and near natural spectrum. The plants undergo primary hardening while they are still developing in the culture.

Such plants have well-developed stems, leaves, and good growth vigour, and survive better on transfer to soil. With proper precautions, contamination of cultures in the bags under greenhouse conditions can be avoided. This system of culturing plants under plastic and greenhouse also eliminates the need, and hence the cost, of air filtration and air conditioning commonly used in the sophisticated facilities. Artificial lights in other laboratory areas such as media preparation, transfer and hardening rooms can also be minimized by incorporating in the room design glass windows that provide indirect sunlight but reduce heat build-up in the rooms. This can be achieved by using either muslin net curtains or translucent glass.

Low Cost Temperature Regulation

Maintaining in vitro cultures at a regulated temperature with air conditioners adds to the cost but does not contribute to specific plant quality. In fact, as in the case of artificial lighting, plants grown under a narrow temperature range are at a disadvantage during hardening and later under the field conditions. Elimination of this factor significantly contributes to reduction in electrical costs. Contrary to the common belief that the day- and night-temperature in the growth room must be strictly controlled at an even level, many in vitro growing plants can tolerate wide fluctuations in temperature. High daytime temperatures of up to 28-30°C and as low as 10-12°C at night do not damage plant growth. On the contrary, fluctuations in temperature promote better growth. In vitro growing plants of banana and potato, kept at 16-41°C at 750 μ mol m^{-2} s^{-1} under natural light, showed as good or better growth than in the controlled growth room.

Reducing Energy Costs for Water and Autoclaving

Normally, distilled water is produced from water stills operated by electricity. Some water stills and autoclaving require a three-phase connection. For small facilities, it is prudent to operate the units on a single-phase electrical connection. In small units, tap water may be used after autoclaving rather than distillation. Pressure cookers heated with gas can also be used where capital is a constraint. In case of large-sized facilities, autoclaves and water distillation equipment operated on electricity is still the most economic.

Reducing Energy in the Glasshouse and Hardening Area

Electrical usage can be reduced significantly from hardening plants in open shade if glasshouses are not available. In vitro grown plants can be hardened by covering them with net-shades after transfer to soil or by placing containers in shade under thatched or plasticcovered open huts. The plants hardened under natural light are sturdy, and withstand transplantation better in the field. In the case of culture in plastic containers, the lids are removed, and containers are covered with a thin plastic sheet, and left for 3-4 days in shade. The medium is then removed by washing with tap water, and the plants are retained for another 3-5 days in the shade before transfer to soil.

REDUCING LABOUR COSTS

Production of plants based on tissue culture technology and their subsequent growing is a labour intensive system. In developed countries, labour is a major cost factor in micropropagation. On the other hand, in developing countries, labour is relatively less costly, which is a major advantage. Yet, in developing countries, increasing the efficiency of production is relevant to reduce the cost of tissue-cultured plants. In developing countries, the access to capital is not easy and borrowing is costly with high interest rates, which adds to cost of production. Hence, it is important to maintain high labour efficiency, especially when the operators in the culture transfer room are not highly trained.

The typical profile of a tissue culture plant production system shows that 40 per cent costs are for labour, 10 per cent for materials, 20 per cent for overheads, and 30 per cent for sales, general and administrative activities. The manual transfer of explants in culture vessels has the upper ceiling of production per operator at about 5000 operations or transfers per day. However, this peak rate is rarely achieved. In general, 2500 transfers per operator per day are considered to be good productivity, which is strongly influenced by the type of culture vessel used. For example, the transfers per person using test tubes are much lower than that in Petri dishes and wide mouth culture vessels.

Once the efficiency of labour in transferring number of propagules per hour has been achieved to the maximum possible, there is little scope to improve efficiency unless automated or semi-automated systems are introduced. These are based on the use of liquid media and bioreactors, and mechanization to handle propagules. Such systems have been developed and can be integrated into the production under certain conditions. Accordingly, such a system can reduce cost of production by 50 per cent in a 20 million-unit production facility. The labour cost comes down from conventional to semiautomated model by 70 per cent and in fully automated model by 75 per cent. The system based on the use of plastic bags increased throughput per person and reduced labour cost per propagule by 60 per cent.

Also, the improvement in contamination control increased multiplication rate and improved survival during hardening and produced more saleable plants. In some cases, the multiplication rate increased from 1.5 to 5, reducing labour cost of tissue culture by 96 per cent per plant before hardening. To reduce labour costs, such systems can be combined so that the multiplication phase is in the bioreactor and the last stages in plastic bags.

To maintain high efficiency of the operators under laminar flow hoods, they should work not more than 4 hours per day and preferably on a single bench; afterwards they should undertake other operations, such as washing, drying of containers or work in glasshouse. The provision of music in culture transfer rooms also increases the efficiency. Usually, female operators are more efficient than males in the transfer rooms and for potting of plants. This contributes to the gender based rural employment.

Reducing Overhead Costs

Overhead costs for commercial scale tissue culture-based plant production include salaries of managers - directors, scientists, marketing managers, market promotional expenses, payment of commissions, and perks to the management. Much of these cannot be avoided. The large-sized companies can afford high overhead costs that match large volume of business. However, smaller units have to rely on a few permanent managers, but may hire the expertise of the specialists as and when required. Large-scale production based on tissue culture is also prone to large-scale losses through contamination of cultures, substandard production, losses during hardening and badly planned and poor marketing.

Many failures of commercial micropropagation concerns can be traced to improper planning of production and marketing. While it is important to maintain high levels of clean technology, the need for critical expertise cannot be ignored in keeping contamination to less than 5 per cent. The high contamination rate can become a major cause of not achieving the set targets and throwing the assumed production of the enterprise out of gear and loss of market.

The advice and expertise of experienced persons is indispensable to reduce such losses. The expertise can be obtained for a reasonable fee either from established R&D laboratories of the larger companies or from colleges and universities. Such laboratories can also provide services for quality control, indexing of mother cultures, advice on media formulations and trouble shooting. For sustainable technology advancement, links should be established within the private sector, and between the private and public sector R&D organizations. The shared costs of R&D in tissue culture technology by the private and public sector is thus a low cost option to the governments and policy makers in the developing countries.

BIOREACTORS AND TISSUE OR ORGAN CULTURE

Bioreactors are vessels designed for large-scale cell, tissue or organ culture in liquid media. For an increasing number of plants, they have demonstrated a number of important advantages over conventional semi-solid micropropagation, including several fold increase in multiplication rates, and reduction in space, energy and labour. These cost-saving advantages have been the driving force for increased attention to the use of liquid systems generally. There are, however, many disadvantages because of a number of problems associated with the use of bioreactors in micropropagation.

These include contamination, lack of protocols and production procedures, increased hyperhydricity, and problems of foaming, shear stress and release of growth inhibiting compounds by the cultures. Because of the inherently difficult problems of contamination in liquid systems, and the lack of problem solving information compared to semi-solid growth media, such systems involve more risk and require greater skill. Thus, in many developing countries, bioreactors have not been a low cost option. The potential of cost reduction will depend on the cost of labour, production capacity, crops being propagated, contamination rate, and energy and costs.

Bioreactor Vessels

For micropropagation, "large-scale" bioreactors generally refer to vessels with volumes ranging from 1.5 to 20, although even larger vessels have been used. However, even these modest volumes are in stark contrast to the standard plant tissue culture vessels, which typically contain less than 0.5l of semi-solid medium. The size of the vessels in bioreactors versus conventional micropropagation systems has important investment cost consequence. Standard semi-solid plant tissue culture vessels have a seal that permits limited gas exchange with minimal contamination. The environment of the conventional growth room determines the light, temperature and gases in the conventional vessel.

Some bioreactors can grow 60-100,000 plants per cubic meter and provide more precise control of the plant growth gaseous exchange, illumination, medium agitation, temperature and pH. Since they are aseptically sealed, they can be operated most of the time in ordinary rooms. Only the handling of plant material is performed under laminar flow. They can therefore be attractive to developing countries as regards new or expanding plant culture facilities, in combination with a conventional laboratory. However, to be cost-effective its use requires indexed plant cultures, and attention to aseptic procedures during the handling of plant material.

TYPES OF BIOREACTORS

Various bioreactor designs have been developed for a wide range of crops, culture types and stages. Functionally, plant culture bioreactors can be divided into two broad types: those in which the cultures are immersed partially or temporarily in the medium, and those in which the cultures are continuously submerged.

The latter are usually used for high-density multiplication of cultures where submersion does not result in abnormal plant development. Such cultures include protocorm-like bodies, embryogenic callus, somatic embryos, meristem clusters and nodules. Protocorm-like bodies are meristematic tissues that regenerate propagules. Meristem clusters and nodules are organogenic culture made up of compact masses of meristems without significant leaf, stem or root development. Submerged bioreactors can also be used for the development of small bulbs or corms for field growing.

Partial Immersion Type

There are several types of partial immersion bioreactors. These include gaseous phase bioreactors, liquid layer bioreactors and temporary immersion bioreactors. Such designs are preferred to culture plant material sensitive to hyperhydricity. In gaseous phase bioreactors, the cultures are mechanically supported on a porous base and intermittently sprayed with medium or exposed to a nutrient mist. Excess medium is collected in the vessel and re-circulated. In such bioreactors the vessels are fabricated from autoclavable clear glass or plastic, and the environment of the growth room controls illumination and temperature. These bioreactors can provide excellent growth and development for most tissue and organ cultures. However the mist generation systems available are complex to maintain.

In liquid layer bioreactors, only the base of cultures is exposed to the medium. The control of illumination, temperature and the gaseous environment is much the same as in standard tissue culture vessels. The 'LifeRaft', a simple liquid layer system, consists of a disposable (or a reusable) microporous membrane raft supported by a buoyant float (reusable, autoclavable) at the surface of the liquid medium. Stationary support systems for liquid layer bioreactors have been also developed from sealed clear plastic film with a wire frame. Small bags made of plastic film are also being used instead of vessels in commercial laboratories.

In temporary immersion bioreactors, the cultures are immersed in the medium, for a preset duration at specified intervals. Their construction and operation is very simple, which makes them attractive low cost alternatives. A typical design, uses two vessels (plastic or glass) of which one holds the liquid medium and the other the cultures. Another version of temporary immersion bioreactors includes a system where a single vessel with a reservoir on one side mechanically tilted at preset intervals. In this manner, medium periodically bathes the cultures in the vessel and maintains the propagules in a vertical position. Other temporary immersion bioreactors use different designed vessels and rotation. As the vessel turns, the culture is intermittently dipped in the medium. Temporary immersion bioreactors are simple and cost effective to run. Consequently, a number of commercial micropropagation laboratories have begun to integrate these systems into their production procedures.

Fig. Rectangular Vessels with Intermittent Cycle of Wetting and Drying on Tilted Shelves

Submerged Type

There are basically two types of submerged bioreactors - mechanically agitated, and airdriven. Mechanically agitated bioreactors make use of devices such as propellers, turbines, impellers, paddles, and vibratory mixers to circulate the medium. Each device creates a somewhat different pattern of fluid flow in the vessel, For each type of culture and plant, the shear force and pattern of flow must be optimized. For example, it is advantageous to have sufficient shear force to maintain meristematic clusters as small aggregates. This minimizes stagnant regions in the vessel, which can cause cultures to stick to surfaces and form clogs that disrupt medium flow. On the other hand, high shear force can produce excessive foaming, and has a negative effect on culture growth. Such a system (e.g. NLH bioreactor) for cell and embryo culture has recently been reported.

Air-driven bioreactors are the simplest submerged bioreactor. In air-driven bioreactors, filter-sterilized air is forced into the vessel's bottom, and aerates and agitates the medium, thereby, lifting and circulating the culture. Such air-driven bioreactors are called simple aeration or bubble column bioreactors, depending on their volume. Air-driven bioreactor vessels are usually fabricated from transparent or translucent materials, but are typically formed from autoclavable clear glass or plastic and fittings. Alternatively, they are made of non-autoclavable clear flexible plastic, and sterilized with gamma radiation.

The simple airlift or bubble column bioreactors are easy and cost efficient to run. The greatest advantage of submerged bioreactors is that they can be integrated into automated or mechanized micropropagation systems. In the automated systems, culture from the bioreactor is pumped to a bioprocessor for separating, sizing and distributing the culture to vessels for propagule development. In the mechanized system, an operator transfers the multiplied culture to a mechanical cutting device for separation and then to a vessel for propagule development.

ADVANTAGES AND DISADVANTAGES

The advantages of bioreactors include increased culture multiplication rate, faster culture growth, reduction in medium cost and reduction in energy, labour and laboratory space. The increased rate of multiplication and growth primarily reflects the effect of liquid medium. The elimination of gelling agents (e.g. agar) reduces medium cost, and filter sterilization of the medium eliminates the need for autoclaving. In bioreactors, the culture density in liquid media is much higher than in the conventional vessels with semisolid media. The conventional tissue culture vessels are typically kept on shelves with a large space between the shelves.

The use of bioreactors requires much smaller space in the growth room, fewer clean work stations, and less space for media preparation, vessel storage and washing than that used in the conventional micropropagation. The smaller size of the laboratory and the number of people reduce air-conditioning needs, hence energy costs. Reduced requirements for lighting the bioreactors and autoclaving, and less labour, simplification of medium preparation, washing of vessels, and simplified handling of the cultures, all lead to cost reduction.

There are, however, many disadvantages because of a number of problems associated with the use of bioreactors in micropropagation. These include contamination, lack of protocols and production procedures, increased hyperhydricity, and problems of foaming, shear stress and release of growth inhibiting compounds by the cultures. Unfortunately, culture contamination, which is a major problem in conventional commercial micropropagation, is even more acute in bioreactors.

In conventional micropropagation, discarding a small number of the contaminated vessels is an acceptable loss; in bioreactors, even a single contaminated unit is a huge loss. However, despite these difficulties, a number of commercial laboratories have developed effective procedures to control contamination in bioreactors.

CONTROL OF CONTAMINATION

Prevention of contamination in bioreactors requires a proper handling of the plant material, of equipment during transfers, and cultures during production. Only the surface sterilized explants, multiplied in small vessels and indexed for freedom from diseases are used to initiate cultures in bioreactors. If the bioreactor is small, it is sterilized in an autoclavable plastic bag, sealed with a cotton wool plug, and opened only under the laminar flow cabinet. If the bioreactor is large, other methods to protect the vessel after autoclaving and during its transport to the sterile area must be provided. The vessel is assembled under laminar flow, taking care that non-sterile objects do not enter the sterile air stream in the vicinity of open ports.

After inoculation, all the ports are sealed and the bioreactor is brought to the growth room and connected to an air supply. For harvesting, the bioreactor is returned to the sterile area, and the ports are opened so that the air stream cannot carry contamination from the outside. If proper procedures are followed, the reactor can be reused without re-sterilization. Despite the precautions taken in initiating cultures, bioreactors can become contaminated from the environment or from microbes, latent in the culture. The contamination can be controlled with one or a combination of anti-microbial compounds, acidification of the media, and micro-filtration of the medium.

Several commercial laboratories use anti-microbial compounds, such as PPM, G-1 (2-(5-bromo-furil)- 1-bromo-1-nitroethane), a fungicidal and bactericidal chemical, sodium hypochlorite or sodium dichloroisocyanurate,. Acidification of the medium (pH 3) has been used to control contamination in the multiplication of banana and in media for other crops. Contamination can also be controlled by circulation of the medium through a 0.2 µ?microporous filter. However, the removal of contaminants needs frequent filter changes, which is too expensive for use in a commercial laboratory. Chilling the filter retards filter clogging significantly.

CONSTRUCTION OF A SIMPLE BIOREACTOR

For laboratories with limited resources, a simple temporary immersion bioreactor can be constructed of autoclavable glass or plastic vessels or bags with a wide (at least 5 cm diameter) mouth and a tight fitting rubber stopper. It is better to use small vessels, of one liter or less. The stopper should have two holes for glass tubing. One tube is inserted and fixed in the stopper so that when the vessel is closed, the end of the tube is closer to the bottom of the vessel. A second tube is inserted and fixed in the stopper so that the end of this tube is near the top of the vessel when it is closed. The tubes can be fixed with a heat resistant silicone sealer, such as that used for repair of car radiators.

The vessel should have at least one port (5- 10 cm diameter) to add or remove cultures, and the port should be sealed with a rubber stopper or a metal closure with a rubber gasket. There are four ports, two for air-inlet and -outlet, and two for medium-inlet and -outlet. The air-inlet and -outlet ports are connected to hydrophobic Teflon submicron microporous air filters, which must be kept dry. The wet filters clog and may lead to contamination. It should be possible to open and close, connect and disconnect medium ports without leakage to prevent entry of contaminants. The bioreactors should fit under the laminar flow unit leaving enough space on the top and sides to allow easy access to the vessel for handling cultures and changing the medium.

The purpose built bioreactors are either air-driven or temporary immersion type. A reservoir vessel similar to the culture vessel is prepared except that the mouth of the vessel is just wide enough to fix the two glass tubes. The tops of the tubes, that reach the base of the culture vessel, are connected to each other with silicone tubing. For both the reservoir and culture vessel, the ends of the short tubes above the stopper are connected to sterilizing filters such as 'Acro 50' filters with a short length of silicone tubing. Prior to use, the bioreactor is assembled with the stoppers loosely fitted into the vessels and placed in an autoclavable bag with a cotton wool plug. The culture vessel can also contain an autoclavable plastic or stainless steel mesh screen to support the culture above the medium in the vessel.

The bioreactor is then autoclaved and placed on the laminar flow bench. The culture is aseptically introduced into the vessel and sterile medium in the reservoir, and the stoppers are tightly closed. If necessary, the part where the stoppers fit into the vessels is taped shut to prevent contamination. The bioreactor is then placed in the growth room, the two vessels are placed alongside each other, and the reservoir air filter is connected to a 6W aquarium pump with silicone tubing. Air is pumped into the reservoir vessel, forcing the medium into the culture vessel. The medium is allowed to remain in the culture vessel for a short duration and the air pump is then connected to the culture vessel air filter, and the medium is pumped back into the reservoir. This process is repeated at preset intervals, and can be automated.A simple bubble reactor has also been described parts of which can be purchased as a kit.

COST-BENEFIT ANALYSIS

A recent study has shown that bioreactors can significantly reduce the production cost of micropropagation in developed countries, from $0.16 to $0.07-$0.08 per unit. This has recently been confirmed in commercial laboratories. For example, a major ornamental plant producer in the US is now propagating 40,000 units per month based on bioreactor multiplication and mechanical culture separation. The cost reduction is based on three situations.

Model 1 is a conventional laboratory producing 20 million units per year. Model 2 is a laboratory using multiplication in bioreactors, mechanical separation of the culture and elongation on semi-solid medium. Model 3 is a laboratory using multiplication in bioreactors, mechanical separation of the culture and elongation in temporary immersion bioreactors.

It is, however, unclear whether bioreactors will produce a commercially relevant cost reduction in developing countries. In countries, such as India, labour cost can be as low as US $0.20-0.60 per hour. Based on projections of Model 1 above, this would mean that in a developing country, a conventional laboratory with 20X106 units per year capacity could produce propagules at $0.06-$0.07/unit. This projection agrees with the available information on commercial laboratories in India with 6-10 million units capacity per year, and production costs of $0.06-$0.14/unit.

For larger laboratories, bioreactor micropropagation could probably further reduce this cost by about $0.01. In addition, bioreactor micropropagation would simplify production management through the use of large culture batches and reduced labour. Although cost savings alone might not justify the integration of the bioreactors into production, the simplification of production procedures and management might make bioreactor-based micropropagation desirable for large-scale micropropagation concerns.

Table: Estimated annual cost (US$) of major investment for the laboratory

Building	Building Costs	Amortize Years	Annual interest %	Annual cost of building	Annual labor cost	Annual energy cost	Total Annual cost
Model 1	2696000	20	4	140190	2163600	209160	2512950
Model 2	2059000	20	4	107070	634800	1374520	879390
Model 3	1641000	20	4	85332	530400	48960	664683

Savings from Model 1-2 = 65 per cent; from Model 1-3 = 73 per cent

Table: Estimated annual cost (US$) of major equipment and furniture for the laboratory

Equipment cost	Furniture cost	Amortize years	Annual interest %	Total annual cost	
Model 1	1000000	300000	10	4	135200
Model 2	100000	30000	10	4	135200
Model 3	100000	30000	10	4	135200

Table: Annual cost of production (US$) for each model (20X106 units/year)

	Total Cost (A)	Total cost (B)	Costs per unit	Unit cost media	Unit cost vessels	Unit cost misc.	Total unit cost
Model 1	2512950	135200	0.13	0.01	0.01	0.01	0.16
Model2	879390	135200	0.05	0.01	0.01	0.01	0.08
Model 3	664683	135200	0.04	0.01	0.01	0.01	0.07

As stated before, the elimination of agar and the use of natural light in the growth room can produce a 90 per cent reduction in resource costs. Similarly, the use of natural light in combination with the use of liquid media can result in over a 90 per cent reduction in the cost production in India. If the adoption of such technologies brings down the cost of production in developing countries, the cost reduction from bioreactor micropropagation would be indeed small.

However, the simplification of production management could still make bioreactor micropropagation worthwhile for the large-scale laboratories. Most laboratories in India (70 per cent) produce less than 1 million units per year. It is likely that this also reflects the situation in many other developing countries. Since the use of bioreactors will not produce a large cost reduction or greatly increase efficiency of small laboratories, the integration of bioreactors into production systems should only be attempted by facilities with skilled and experienced propagators.

PRIMING TISSUE CULTURED PROPAGULES

Priming of in vitro propagules refers to the manipulation of the growing environment, prior to and upon transplanting, and is an integral part of tissue culture propagation. Tissue-cultured propagules are produced under controlled environment. Such plantlets have small juvenile leaves with reduced photosynthetic capacity, and malfunctioning stomata. Priming for rooting, shooting, and improved photosynthesis can be achieved with growth regulators and simple adjustment to the growing conditions that affect the post-transplanting performance of the propagules.

Vented closures with microbial filters have been used to facilitate gas exchange to reduce ethylene build-up that stunts plant growth, reduces leaf-size, and causes leaf drop in tissue culture containers. Plantlets produced under photoautotrophic culture systems on media with or without sucrose but CO_2-enrichment, increased light intensity, good gas exchange and reduced humidity are more vigorous, have larger root- systems, and are less susceptible to microbial contamination. Plants adapted gradually to the ex-vitro environment have improved survival upon transfer to soil. Plants in their natural environment are colonized with many bacteria, fungi, and mycorrhizae. In-vitro or ex-vitro biopriming of micropropagated plants with such organisms can improve plant performance under stress environments, and consequently enhances yield.

The adaptation of plants to the environment - temperature, osmotic stress, and pathogens, is triggered by modification of expression of specific genes and metabolic pathways. Certain chemicals, environmental factors, and microorganisms pre-sensitise the cellular metabolism of plants. Upon exposure to stress, the pre-sensitised or primed plants adapt better and faster than the non-primed plants. The production of high quality and vigorous plants through in vitro culture requires the enhancement of post-transplanting ability for water management, efficiency of photosynthesis and resistance to diseases. Priming of in vitro propagules, based on manipulation of the growing environment (chemical, physical and biological) prior to and upon transplanting, is an integral part of tissue culture propagation. The ability of the propagules to withstand transplanting stress very often determines the success or failure of tissue culture operations.

Tissue-cultured propagules are produced in controlled environments under pre-set temperature and low-light intensity, and in small vessels with poor air exchange. This leads to high humidity and accumulation of ethylene. Also, the nutrients in the growing media, sucrose, mineral salts, vitamins and growth regulators, are far in excess of those in the soil. Such an environment causes developmental distortions, and represses or modulates several metabolic pathways. As a result, the plantlets have small juvenile leaves with reduced photosynthetic capacity, and malfunctioning stomata. The roots of such plants lack hair or have few, and the leaves have poor cuticle development and low wax deposit.

Priming for Rooting

The ability to manage water, photosynthesis and response to stress during and after in vitro culture determines the final performance of the propagules. The formation of a functional root system is an essential step to manage water under the post vitro conditions. The in vitro roots remain functional and continue to grow during ex-vitro acclimatisation. Easy to root species, (e.g. potato) usually regenerate roots upon subculture on hormone-free medium.

However, root architecture and biomass partitioning between roots and shoots can be easily manipulated by simple adjustment to the growing conditions that affect the post-transplanting performance of the propagules. For example, increasing day/night temperature from 20/15 to 33/25°C either slightly stimulates or has no effect on the shoot growth in clones of potato, but dramatically enhances root growth and root branching.

Several chemicals also influence the development of the root system. Abscisic acid is known to affect initiation of new lateral roots and overall changes to the root architecture Jasmonic acid (JA) has a similar effect. Low JA concentrations (0.5- 1.0 μ M) stimulate both root and shoot growth. Jasmonates induce stress resistance in plants, and may be useful for priming of in vitro plants during the last subculture.

In potato, nodal explants from JA primed stock plants tuberize earlier and more uniformly, and produce more microtubers than the non-primed controls. Salicylic acid (SA) is a major signalling compound and induces abiotic and biotic stress resistance in plants. It also induces adventitious roots in cuttings and stomata closure, and inhibits ethylene biosynthesis in detached leaves; hence it may also be useful for priming tissue-cultured propagules. Many woody and herbaceous perennials require auxins for rooting. Auxins are involved in root initiation and adventitious root formation. However, excessive auxins after root induction may inhibit root regeneration and elongation, and impair formation of a good root system.

Therefore, a careful selection of the auxin type and concentration is recommended. IAA and IBA are usually used for easier-to-root plants and NAA for more recalcitrant woody plants. The efficacy of different auxins also depends on the explant type, and exposure to light. For example, a short duration (3-7 days) of dark incubation enhanced rooting in apple, and papaya. Addition of low concentration of the growth retardant daminozide to hormone-free medium enhanced post-transplanting acclimatisation of potato plants. Other supplements like activated charcoal, seaweed concentrate, pyroligneus acid, and dilution of salts in the basic MS medium can also aid root initiation and growth. In vitro cultured papaya shoots in aerated vermiculite gave better rooting with higher post-transplanting survival than on non-aerated vermiculite and non-aerated agar.

In some difficult-to-root species, e.g. apple and Saskatoon berries, in vitro rejuvenation via several subcultures is required before successful rooting of microcuttings. Individual cultivars may require adjustment to culture conditions and different number of subcultures. For example, over 90 per cent rooting was achieved with apple microshoots, cv. 'Jonathan' by the ninth subculture, but a more recalcitrant variety 'Red Delicious' required over 30 subcultures for successful rooting. In Saskatoon berries, where rooting by traditional stem cuttings is very difficult, tissue culture rejuvenation and subsequent transplanting to the greenhouse can generate highly vigorous plants. Shoot cuttings taken from such planted hedges readily root within 2-3 weeks after a simple treatment with IBA.

Fig. Five-week-old Potato Plants cv. 'Kennebec' Grown on 10 ml Hormone-Free MS medium (pH 5.7, 3 per cent Sucrose, Vitamins and 0.6 per cent agar), Supplemented with various Concentrations of JA at 22/18ºC day/night Temperature, 16 h Photoperiod, 150 μ E-2s^{-1} Fluorescent Light

Increasing Photosynthetic Competence

Tissue culture containers provide sterile conditions and high moisture, thereby preventing contamination and desiccation of the explant, but restrict aeration and escape of metabolically generated ethylene and water vapors. Such conditions slow down growth and induce undesirable morphological changes. The ethylene build-up stunts plant growth and reduces leaf-size, and causes leaf drop. Such plants desiccate quickly upon transfer to the external environment. Vented closures, including those for baby-food-jars with built-in different sizes of microbial filters, have been used to facilitate gas exchange between culture vessels and the ambient environment.

High relative humidity in culture vessel reduces deposition of epicuticular waxes on leaves. It impairs the functioning of stomata and alters leaf mesophyll by reducing the number of palisade cells and by forming spongy parenchyma with large air spaces. Other effects include low nutrient uptake due to decreased transpiration, decreased lignification, and reduced or abnormal development of the vascular system.

Although the cuticular wax development and stomata functioning can be induced by gradual adjustment to lower humidity, morphological changes cannot be easily reversed. The altered leaf morphology reduces photosynthesis and enhances respiration. Cooling the bottom of culture vessels by 2-3°C below the ambient air temperature in the culture room or providing high light intensity can reduce high humidity. Gradual opening of vessels for a few days prior to transplanting can also adjust high humidity.

Many micropropagated plants have C3 photosynthetic pathway; hence, are greatly affected in their ability to assimilate carbon. The balance between photosynthesis and photorespiration controls their productivity. Under the traditional methods of micropropagation, cultures are usually mixotrophic (use carbon source mostly from sucrose in the medium as well as photosynthesis).

The carbohydrates in the media inhibit carbon assimilation via photosynthesis by reducing rubisco (ribulose-1-5-bisphosphate carboxylase/ oxygenase) activity. On the other hand, several studies indicate a stimulatory effect of sugar on biomass accumulation under elevated CO_2.

In the photoautotrophic culture systems the cultures are grown under CO_2-enrichment (1000-1500 μ mol mol^{-1}), increased light intensity (at least 150 μ mol m^{-2}, s^{-1}), good gas exchange and reduced humidity, and on media with or without or sucrose. Plantlets produced under such conditions are more vigorous, have larger root- systems, and are less susceptible to microbial contamination. They are able to photosynthesize, and do not require acclimatization during post-vitro establishment.

Priming for Temperature Tolerance

In vitro grown plantlets can be acclimatized to low temperatures for cold tolerance under field conditions. Alfalfa and red clover seedlings withstood -16 and -10° C, respectively, upon cold hardening at 2-5° C, 80ìE m^{-2}, s^{-1} light and 8h photoperiod, for 3-5 weeks. Such hardening has been successfully utilized in the selection of cold tolerant alfalfa and red clover genotypes. Elevation of sucrose in the culture medium or application of abscisic acid, or both, may further improve plantlet hardiness in some species.

Acclimatization to Soil Transfer

Good quality propagules with well developed roots and leaves are easy to acclimatize to the external environment. Any successful acclimatization protocol must ensure that the plants maintain active growth during the entire weaning period. The effect of in vitro growth conditions on post-vitro acclimatization has been demonstrated in grapevine and chile ancho pepper, Capsicum annuum. The most common approach to improve plant survival upon transfer to soil is their gradual adaptation to the ex-vitro environment. Under such conditions, plants rapidly convert from heterotrophic or photomixotrophic to autotrophic growth, form fully functional root systems, and their stomatal and cuticular transpiration is reduced.

Such gradual adaptation is often carried out in greenhouses by decreasing relative humidity using fog or mist chambers and increasing light intensity and shading techniques. Light intensity needs to be adjusted to the plant requirements. Rooting of in vitro produced shoots in vivo is recommended for some species to promote development of roots already adapted to external conditions, to prevent potential damage to fragile in vitro roots, or to avoid post-rooting dormancy. Other treatments of in vitro plantlets upon transplanting to the outside environment include high phosphate fertilizer to enhance their vigour, fungicides to prevent damping-off and anti-transpirants to prevent desiccation; however, they may cause phytotoxicity.

A simple, low cost method for ex vitro weaning and liner production of woody plant species and Hosta has been developed. The shoots with root-initials from the rooting medium are transplanted in 'Jiffy' plugs. Plants are acclimatized in plastic trays (25x50x5cm) with transparent cover. Each tray holds 300 plugs, each 18mm diameter. The plugs are saturated in a 2.5-3L solution of fertilizer (1g/L) 10-52-10 NPK; the excess solution is discarded before transplanting. The plants are watered weekly (approx. 1L water/week) and kept in the growth room benches for 3-weeks after transplanting at 22-24/20ºC day/night, 14h photoperiod, under 120 μ $Em^{-2}s^{-1}$ fluorescent light. The current cost is US$73 for 5400 plugs, $58 for 100 trays, and $60 for 100 plastic covers; the total cost per plug being US$0.018. The rooting success in woody plants, e.g. Swedish columnar aspen, Saskatoon berry, chokecherry and pincherry cultivars, sour cherry, low bush blueberries and others was more than 95 per cent, and 100 per cent with hosta varieties.

In seed potato production, 2 to 4-week old tissue cultured plantlets are usually hardened off for 2-4 weeks in a greenhouse, and transplanted in the field. Such plantlets are still juvenile and not well adapted to stress. To prevent desiccation, insect and bird damage, and spread of viruses by aphid, many growers cover plants with floating covers and install an overhead irrigation system. Seed potatoes produced under floating covers are consistently better in quality than those from the unprotected planting.

Fig. Plant acclimatization in plastic trays with transparent cover. Left. Trays in a growth room bench next to culture vessels. Right. Aspen (Populus tremula var. erecta) liners in Jiffy plugs, three weeks after transplanting

In vitro produced-propagules of woody species, e.g. Saskatoon berry (Amelanchier alnifolia Nutt.), Chokecherry (Prunus virginiana L.), Pincherry (P. pennsylvanica L.), Swedish columnar aspen (Populus tremula var. erecta), European weeping birch (Betula pendula var. 'Gracilis' become dormant after rooting. In vitro rooted Saskatoon berry is particularly prone to dormancy as stress factors readily induce terminal bud dormancy. When the shoots stop growth, leaf shedding occurs in the culture or soon after transplanting, and the propagules desiccate. The liners produced from such plants are less vigorous and difficult to establish in the field.

In Chokecherry (P. virginiana L.), dormancy symptoms of shoots with reddish leaves and cessation of terminal-bud growth are induced upon depletion of nutrients in the medium. Post-rooting dormancy has also been reported in rhododendrons and low bush blueberries. Adding low amounts of gibberellic acid (GA) to the medium prevents dormancy, although shoots treated with GA are difficult to root. In Saskatoon berry, ex-vitro rooting under non-sterile conditions is helpful, although does not eliminate the problem. Spraying leaves of ex vitro rooted-shoots with BAP (400ppm), followed by GA (100-PPM) the next day prevents dormancy.

Priming Somatic Embryo-Derived Plants

Somatic embryos are produced as adventitious structures directly on explants of zygotic embryos, from callus and suspension cultures. The stress inducing molecules in plants, ABA and JA have been successfully used to facilitate embryo maturation and priming for resistance to post transplanting stress. ABA and osmotic shock with 6 per cent sucrose is a routine step to induce desiccation tolerance in alfalfa synthetic seeds.

Osmotic stress induction with 3 per cent sorbitol was effective in embryo maturation and conversion in soybean cultures and ABA treatment combined with partial desiccation of encapsulated somatic embryos in sugarcane. Several sugar alcohols and polyethylene glycol have been tested with different degrees of success for priming SE propagules of various plants. In conifers, post ABA embryo maturation has been significantly improved by benzyl adenine (BA). The lowering of sucrose concentration from 58 to 29mM in the proliferation medium improved the formation of cotyledonary SE on the maturation medium. Interaction between plant growth regulators and various carbohydrates and organic nitrogen compounds in embryo development and maturation has been reported.

BIOPRIMING

Plants in their natural environment are colonized both by external and internal microorganisms. Some microorganisms, particularly beneficial bacteria and fungi, can improve plant performance under stress environments, and consequently enhance yield. Plants infected by microorganisms develop systemic resistance (systemic acquired resistance, SAR, or induced systemic resistance, ISR), and/or benefit from their antagonistic abilities towards pathogens (cross protection).

Although, the inoculation of seeds with beneficial microorganisms has been practiced for more than 50 years, the inoculation of tissue culture propagules to enhance plant performance is relatively new. Plant tissue culture is based on axenic (contaminant-free) culture systems. Hence, microorganisms, including endophytes, are treated as problemcausing contaminants, and various procedures have been developed to eliminate them. Only recently, have microbial inoculants, primarily bacterial and mycorrhizal, been evaluated as propagule priming agents both as in vitro co-cultures and on transplanting.

Standard microbiological techniques allow culturing only a few per cent of the naturally occurring microorganisms. Moreover, the soil and plant associated bacteria can switch between culturable and unculturable stages. The development of new culture methods that allow the establishment of stable associations between plants and beneficial organisms, both in vitro and ex vitro, under different environments is a major challenge of future research.

In Vitro Bacterization

Rhizosphere bacteria interact with plants and other inhabiting organisms by producing a wide array of chemicals. Many of these chemicals exhibit plant regulatory activity. There is increasing evidence that the bacteria associated with plants (endophytic and epiphytic) induce host resistance to environmental stresses. This is commonly linked to the production of salicilate or jasmonate, or both. These two types of signalling chemicals are associated with resistance to biotic stress.

Recent molecular evidence in Arabidopsis thaliana inoculated with the plant growth promoting rhizobacterium, Paenibacillus polymyxa, indicates however, that the abiotic and biotic stress-resistance are linked. An effective plant beneficial bacterium, Pseudomonas spp. strain PsJN, was used to enhance tolerance to transplanting stress in potato. The bacterium was originally isolated as a contaminant from Glomus vesiculiferum infected onion roots. A key element in its isolation was that the bacterium (Pseudomonas sp. strain PsJN) did not grow on a standard potato culture medium in the absence of plantlets. Pre-selection of plant beneficial bacteria on plant tissue culture media is thus recommended as a critical step in the development of an in vitro co-culture system, so the bacteria do not overgrow the culture.

Dipping potato nodal cuttings in a suspension of a beneficial bacterium can dramatically affect in vitro growth of plantlets. So far, our isolate, Pseudomonas sp. strain PsJN, has been the most effective plant growth promoting bacterium under in vitro conditions. It forms endophytic and epiphytic populations when co-cultured either with potato, tomato and grapevine. Thus, in clonal propagation by nodal explants, there is no need for further re-inoculation. The bacterium stimulates plant growth and induces changes that lead to better water management. It also enhances resistance to low level pathogens of potato and tomato and to Botrytis cinerea infection of grapevine.

The inoculated plantlets had a massive and wellbranched root system after four weeks in culture, and were more advanced developmentally. The stems were sturdier with more lignin deposits around the vascular system, more root hairs and more and larger leaf hairs. Stomata function of the bacterized in vitro plantlets closely resembled those of the greenhouse hardened non-bacterized transplants. In wheat seedlings exposed to osmotic stress, improvement of water relations has been observed by their co-culture with Azospirillum brasiliense strain SP245. Enhanced transplant survival of black locust, inoculated with different Rhizobium isolates selected for growth vigour from Rabinia varieties, has also been reported.

The stimulation of adventitious rooting in cultures with Agrobacterium rhizogenes is well known. It was successfully used to enhance rooting of two Mexican species of Pinus. Rooting of somatic embryo-derived adventitious shoots (15-30 mm long) increased between 7-13 per cent, and in shoots co-cultured with A. rhizogenes to 67 per cent compared with 60 per cent in the auxin treated controls. Root development was also induced in slash pine explants on co-culture with the bacterium isolates. The bacterium-induced roots resembled seedling roots rather than hairy roots, which are induced with A. rhizogenes.

Induction of root elongation has been observed in potato plants co-cultured in vitro with some strains of Rhizobium leguminosarum biovar trifolii, isolated from the root nodules of red clover. The bacterization of shoot explants of oregano with a Pseudomonas sp. prevented vitrification. The oregano plantlets co-cultured with the bacterium had lower water content, but more phenolics and chlorophyll than the non-bacterized controls. Similar responses were recorded with potato and vegetable crops. Bacterized plantlets were greener, had elevated levels of cytokinins, PAL (phenylalanine ammonia-lyase), free phenolics (especially chlorogenic and caffeic acids), and contained more lignin. In potato, elevated PAL activity in bacterized plants was found during the first four weeks of growth, there being no difference in its level in six week-old plants.

However, when explants were taken from six-week-old bacterized and control stock plants, PAL activity in the newly derived shoots was higher in the pre-bacterized treatment. This suggests that pre-bacterized plantlets respond faster to cutting injuries, a critical factor in the induced host resistance to pathogen attack. In vitro bacterization of potato and vegetable plantlets significantly enhanced their post transplanting survival vigour. Enhanced vigour of potato plantlets co-cultured with isolates of Pseudomonas fluorescence isolated from tubers, during and after weaning has also been reported.

In greenhouse experiments, plants derived from dual cultures of potato and our pseudomonad bacterium had larger root systems and stolons, tuberized earlier, and gave higher tuber yields than the non-bacterized controls. The bacterized potato plantlets transplanted directly from culture vessels to the field had significantly better survival than the non-bacterized controls; however, the tuber yield varied from year to year; high precipitation or severe drought reduced the benefits of bacterization.

Tissue culture techniques provide an opportunity for the introduction of nitrogen fixing endophytes (other than Rhizobia) into clonally propagated plants for sustainable production systems. A protocol has been developed to inoculate micropropagated sugarcane plantlets with Acetobacter diazotrophicus at the end of the rooting period on medium with 10X diluted salts and sugar but without hormones and vitamins. Thirty days after transplanting, the bacteria could be re-isolated from the plant tissues. Stable artificial associations between plants and nitrogen fixing bacteria have been also reported.

A symbiotic culture system of callus and bacteria between Daucus carota L. and Azotobacter zettuovii (CRS-H6) was developed. The callus grew for four years on the nitrogen-free medium with lactose as the carbon source. The bacteria located in the intracellular spaces were transmitted to newly regenerated plantlets and fixed nitrogen. Similar stable associations were established with tomato, potato, wheat, sugarcane and poplar, using different strains of eight Azotobacter species (Varga, S.S. Personal communication), and in strawberries with Azomonas insignis. In tomato, the bacterium was transmitted through the seed. Approximately, 20 per cent of the seedlings from the seed of tomato cocultured with Azotobacter beijerinckii showed N2 fixing activity two months after germination.

Ex Vitro Bacterization

To comply with the certification requirements of plant tissue culture propagules, commercial laboratories are hesitant to introduce microorganisms into their in vitro cultures. One approach is to wean plants in rhizosphere with both beneficial bacteria and mycorrhizal fungi. In potato, the tissue-cultured plantlets are readily infected both in external and internal tissues with bacteria present in the transplanting medium. This approach, although attractive, will have to deal with the problem of maintaining stable populations of the introduced microorganisms.

The beneficial effects of bacterial inoculation on tree seedlings and greenhouse produced potato tubers show the potential of post-vitro bacterization of tissue-cultured propagules. The post-vitro mycorrhization and bacterization of micropropagated strawberry, potato and azalea with certain combinations of bacteria and mycorrhiza enhanced greenhouse production of minitubers, and a mixture of three strains of rhizobacteria improved the post-transplanting performance of strawberries.

Mycorrhization

The inoculation of tissue-cultured plantlets with endomycorrhizal fungi is beneficial to plant growth. It is well known that the root colonization with vesicular-arbuscular mycorrhizae (VAM) improves plant nutritional status, water management and disease resistance. Several secondary metabolites (cyclohexenone derivatives) with structural similarity to abscisic acid have been identified in tobacco roots in response to the Glomus intraradices infection. A variety of phenylpropanoid compounds induced by ectomycorrhizal inoculants, have also been identified in several conifers. Such compounds are related to the biotic and abiotic stress management in plants.

To improve the performance of tissue-cultured plantlets, rooted-shoots have been inoculated with mycorrhiza at transplanting. The benefits of mycorrhization depended on the growing medium, plant and mycorrhizal species, and the degree of root colonization. Because VAM stimulated growth of plants was frequently expressed only after acclimatization, it was suggested to develop a culture system where the mycorrhizal fungi are introduced in vitro during the rooting stage.

It is possible to grow fungi in vitro on Ri T-DNA transformed carrot roots under elevated CO_2. Capitalizing on this technique, a 'tripartite' culture system for in vitro inoculation of strawberries and vegetable crops was developed. The most effective treatment consisted of 30 day-old VAM transformed carrot-root culture and strawberry shoots in cellulose plugs. After root induction, the plugs were placed on the surface of the mycorrhized root culture and kept in a growth chamber under 5000 PPM CO_2 for 20 days.

All plantlets were successfully colonized, and exhibited a larger root system, better shoot growth, and a higher (more negative) osmotic potential when compared to non-mycorrhized controls. It is suggested that the enhancement of osmotic potential is important in the pre-adaptation step prior to full acclimation of plantlets for transplanting. The high osmotic potential of wheat seedlings co-cultured with Azospirillum allowed them to withstand osmotic stress much better than non-inoculated controls with low osmotic potential.

In strawberry, mycorrhized plants had a better establishment rate and produced more runners than non-mycorrhized controls. The plants were grown in polyurethane foam substrate under photoautotrophic conditions with reduced sucrose concentration in the medium. The foam tear-away strips fit into any culture vessel, and the VAM spores can be placed directly in the planting holes. In garlic, improved growth was observed after post-vitro transplant inoculation. Increased rooting and reduction in weaning stress have been reported in the medicinal plant, Babtista tinctoria (L.) R. BR..

Several studies have demonstrated synergistic effects of VAM fungi and diazotrophic bacteria on nutrition and growth of various crops. Such effects of "mycorrhiza helper bacteria" need to be explored in tissue culture systems. Creation of defined tissue culture micro-ecosystems could allow the study of complexed plant–microbial-environment interactions, which could refine and improve our traditional in vitro propagation methods and prime the propagules for ex-vitro environments.

Priming with Microbial Derivatives

Cyanobacterial and fungal culture extracts have been shown to induce developmental changes and pathogen resistance in tissue-cultured material when added to the medium. Microbial culture filtrates have been reported for the control of plant tissue culture contaminants. An acetone precipitated fraction of Bacillus subtilis Ehrnberg and Trichoderma viride Pers exhibited high antifungal activity, and extracts from Pseudomonas fluorescence Migula (strain X) had high antibacterial activity.

The bacterial extracts did not reduce the performance of cultures of Nicotiana tabacum L. over four subcultures but reduced the growth of accidental contaminants. Heat-stable mycelial extracts of another Trichoderma species, a non-pathogenic T. longibrachiatum were triggered induced-resistance response to Phytophthora parasitica var. nicotianae (race 0) in tobacco seedlings. The extract induced expression of pathogenesis-related proteins, PR-16 and osmotin (PR-5) at a higher level than the extract prepared in the same manner from the pathogenic fungus.

LASER APPLICATION

Using a special single chamber it was possible to perform laser photocoagulation to the RPE by positioning the chamber perpendicularly to the laser beam. Laser treatment spots (distance 200m) per tissue ring system was performed one day after cultivation using a conventional argon laser [wavelength: 514 nm, pulse duration: 100 ms, spot size: 200 μm (even on tissue), power: 400–1,000 mW]. Different energy levels were utilized according to different pigmentation of the tissue.

Thus energy was chosen in a way that a laser effect could be recognized ophthalmoscopically as a gray lesion. Also noticed was some pigment which was released from the RPE during the laser application and always a centrifugal enlarging lightening of the RPE could be observed in the first milliseconds after irradiation.

The lesion appearance varied according to the laser power, however, the overall variability of the ophthalmoscopic aspect of the lesions was considered to be minor. The irradiated tissue was kept under perfusion conditions for 1, 3, 7 and 14 days and examined histologically afterwards by light microscopy.

STUDY DESIGN

The specimens were divided into two subgroups: perfused RPE without laser treatment and perfused RPE with laser treatment. Each laser-treated specimen received nine laser spots having a distance to each other of about 200 μm.

LIGHT MICROSCOPY

Tissue fixation was performed in formalin 10 per cent for paraffin embedding to get a first impression and then in 3 per cent glutaraldehyde (Sigma) in DMEM without porcine serum. Fixation took place after about 24 h. Specimens then were embedded with EPON.

Perfused RPE without Laser Application

Under perfusion untreated specimens were well differentiated. Regular RPE formation with apical melanin was seen 1 day and 3 days after cultivation. The RPE cells were in close apposition and some microvilli at the surface of the RPE layer were recognized. After 7 days the RPE formation changed. Apical melanin was still present but marked alterations of the RPE cells occurred yielding a cuboidal and partly dome shaped morphology.

Some vacuoles could be found inside the cell suggesting phagocytosis. After 14 days of perfusion culture the RPE was multilayered. After 28 days further growth of the multilayered RPE was noticed in one specimen. In all histologic sections the choroid showed a good preservation of the matrix with well differentiated vessels having an intact endothelium.

Perfused RPE with Laser Application

It was difficult to identify suspected areas as proper laser lesions. In all specimens examined only a minority of the laser spots applied were found by sectioning the tissue. Thus many areas which were suspected to be laser lesions did not show cell proliferation a few days after irradiation; however, in these cases differentiation from artifacts by preparation was difficult. Thus for all given times of examination just six laser lesions could be clearly identified as areas treated by laser having following appearances: One day after laser treatment the irradiated area could be identified as an area devoid of RPE cells.

Bruch's membrane remained intact judged by light microscopy but appeared wrinkled over the whole length of the laser burn. The underlying choroid had closed vessels, extra cellular pigment and cell debris. After 3 days, migration and proliferation of cells derived from the RPE next to the lesion could be noticed. These cells were flat and revealed some melanin granules. Migration took place towards the defect in the RPE layer. At the edge of the lesion Bruch's membrane was wrinkled and choriocapillary vessels were closed.

After 7 days, proliferation could be clearly demonstrated. The whole defect in the RPE layer due to the laser impact was already covered by a mound of focally nodular, multilayered RPE. Maximal proliferation was found predominantly at the edge of the lesion. In this area the melanin granules lost their apical orientation. Bruch's membrane was strongly wrinkled in that area. The choriocapillary vessels were open at the edge of the burn.

The whole lesion was estimated to be 340 Ìm in diameter, but proliferation made it difficult to determine the exact dimension of the lesion. Hence due to the cell proliferation of untreated cultured RPE it was also difficult to determine the diameter of the laser lesions 7 days after cultivation. After 14 days, cells with broadly distributed melanin granules were present. Multilayered cells showed some prominent proliferations. The diameter of the lesion was not determinable because of the general cell proliferation but Bruch's membrane which was focally wrinkled again, was indicating the presence of a lesion.

Control Experiment

The permanent perfused culture model offers the possibility to cultivate retinal tissue in an organotypical environment having a perfusion line from both sides, the upper RPE and the lower choroid side. Using this RPE organ model in perfusion culture, it was possible to examine the behaviour of the RPE after laser photocoagulation and the behaviour of the organ culture without treatment.

The proliferative behaviour of the RPE is associated with the substratum on which the cells are cultured. We used the perfusion culture system with the RPE monolayer still attached to Bruch's membrane. In previous static culture approaches it was difficult to maintain the RPE as an intact monolayer. Numerous workers developed culture models but had difficulties finding a solid support for the exoplants of human RPE and choroid.

Tso et al. used filter paper (Millipore) as support for the isolated uveal tissue and noted a markedly wrinkled Bruch's membrane and hypopigmented RPE cells after only 3 days in organ culture and frequent focal nodular proliferation within 10 days of culture. In contrast Del Priore and coworkers did not observe these alterations using another model. They were able to maintain the RPE as a monolayer for some weeks and explained this by the fact that the RPE and choroid remained attached to the structural support of the underlying sclera in their organ culture system. In their experiments the RPE monolayer remained intact after 2 weeks but they also noticed hypertrophic RPE cells and areas of RPE hyperplasia.

Our approach using the permanent perfused culture system supports the findings of other authors, who recognized proliferation of the cells. It was speculated that with this new culture model, RPE without neurosensory retina should be maintained as an intact monolayer for longer than 2 weeks. After 3 days in the perfusion culture the RPE monolayer was intact, after 1 week the cells were also intact as a complete monolayer, though there were some morphologic changes.

However, after 2 weeks a marked proliferation of the cells and multilayered RPE was observed. Proliferation of the cells as seen beyond 7 days could be interpreted as a sign supporting the vitality of the organ culture. However, the morphologic findings after 14 days of culture may also suggest a phenotypical change of the RPE cells into other cells such as macrophages. Alternatively an overgrowth of the cell layer due to choroidal cells seems to be unlikely since Bruch's membrane stayed intact.

In our study these cell alterations of untreated RPE limits the use of the culture system for RPE organ culture over a period longer than 2 weeks. It is not clear why this cell change occurred in the RPE culture. When full thickness adult retina is cultured in the perfusion system, no signs of cell changes are observed, and the RPE remains intact as a monolayer over a period of 4 weeks. Therefore it could be speculated that with the loss of neurosensory retina the RPE is no longer inhibited, and thus reacts with cell proliferation and subsequently further scarring. In comparison to the static culture approach as described by Del Priore the perfusion culture system revealed no advantage for RPE cultures since proliferation of untreated cells was evident after 7 days of culture, whereas Del Priore was able to maintain the RPE for 28 days.

Laser Application

The effects of laser treatment of the fundus have been studied by several groups. In vivo it has been observed that argon laser photocoagulation of monkey and human fundus causes necrosis of the RPE and a lifting of the RPE from Bruch's membrane, budding of individual RPE cells and multilayered RPE formation in the area of laser irradiation by 7 days after treatment. Histologic sections revealed that by irradiating the RPE with a conventional argon laser the whole area of the cells was destroyed and the choriocapillaris as well as the choroidal vessels were damaged.

After laser photocoagulation RPE cells migrate and proliferate to cover the defect. In vivo after mild photocoagulation (just light whitening of the irradiated area), as usually performed in macular coagulation, the RPE barrier becomes intact again.

Histological examinations of rabbit retina showed also a proliferation of the RPE after coagulation with short pulsed repetitive laser systems without destruction of the neurosensory retinal layer. So RPE proliferation can always be observed as a healing mechanism to repair any defects in the RPE monolayer or adjacent structures. The effects described could also be demonstrated in the laser lesions of our approach investigating mild RPE laser photocoagulation in vitro. In these examples, as in vivo, RPE cells started to cover the defect by migration and proliferation.

After one week the defect was completely covered by multilayered RPE cells showing still an ongoing proliferation after 14 days. Similar to the findings of Del Priore, one day after argon laser photocoagulation we could also observe a release of the RPE from Bruch's membrane and a total destruction of the irradiated area. It was interesting to note that Bruch's membrane itself was not destroyed but appeared wrinkled.

Thus it was possible to show in vitro the migration and proliferation of the neighboring RPE cells to cover the laser defect up to 7 days. This is consistent with the in vivo findings of the RPE selective laser treatment showing that the RPE is replacing the irradiated area. These healing mechanisms occur even without overlying neurosensory retina. This is an important observation because the neurosensory retina contains potent growth factors that could affect the morphologic response of the RPE to laser photocoagulation. However, since this is not an audio-radiographic study, it should be noted that the 'proliferation' might just be a metaplasia rather than a mitosis of the cells.

Limitations of the Study

Most authors did not give information about the number of laser spots examined, sizes of lesions in histology and percentage of proliferations. In our preliminary experiments the described effects of laser-tissue interaction could be observed in a minority of all applied laser spots. This could be mainly explained by the difficulties in identifying the laser burns when trimming the tissue due to the fact that no thermal necrosis could be seen because of the lack of neurosensory tissue. Thus major damage is only confined to the RPE and in spite of the macroscopically greyish appearance of the RPE in irradiated areas histological alterations are hard to identify. Concerning acute laser lesions Bruch's membrane is denuded and void of RPE cells.

In our study several areas without RPE were found – also 3 and more days after irradiation – but differentiation from artifacts due to tissue preparation was difficult and therefore no fraction of proliferating lesions could be mentioned. In contrast to this, localized proliferations in treated specimens were a clear sign for laser lesions. Another explanation for the irregularity of present proliferations could be the possible variation in laser energy, since it turned out to be difficult to judge experimentally an endpoint of laser treatment, when neural retina is absent. However, in all treated specimens laser burns were slightly visible indicating a destruction of the RPE which should be independent from the delivered energy.

Also the variation in spot size due to different focussing could be a contributing factor for the different proliferation behaviour of the RPE. Interestingly the described laser lesions showed variable sizes of destruction. Although the chosen diameter of irradiation was about 200 Ìm, the diameter of the histologically observed laser lesions varied. There are several explanations for these findings. The focus during laser application through the microscopic culture chamber could change the diameter and additionally, each burn was adjusted and focused individually with a certain variation. Secondly, RPE cells responded variably to laser coagulation depending on their melanin content. This would lead to a variation in heat conduction. After 7 days it was nearly impossible to determine the exact size of the lesion because of the overall proliferation.

A major limitation for studying laser-tissue interaction for periods longer than 14 days seems to be the perfusion culture system itself because both treated and untreated tissue revealed proliferation and thus differentiation and demarcation of treatment results are difficult. Therefore, further research should consider improving the perfusion system, e.g. by changing the buffer solution or the perfusion time. This model is useful to study laser effects of the RPE in vitro although preparation and histological documentation of laser spots can be difficult. However a healing process of the RPE could be observed associated with migration and proliferation of the cells. On the other hand, untreated RPE also showed a proliferation after 7 days in the perfusion culture. After 2 weeks a general alteration of RPE cells to multilayered cells occurred. This morphogenesis of cultured RPE cells has been already shown by several authors.

As a consequence of these RPE alterations our system seems not to be useful after a period of 7 days to examine laser tissue interaction. In contrast, it might be suitable – because of short-term changes – to perform e.g. toxicity tests with intraoperatively used balanced solutions or investigations on growth factor production or RPE transplantation in the early period.

2

Tissue Engineering, Stem Cells and Cloning

INTRODUCTION TO TISSUE ENGINEERING

As one of the major components of regenerative medicine, tissue engineering follows the principles of cell transplantation, materials science, and engineering toward the development of biologic substitutes that can restore and maintain normal function. Tissue engineering strategies generally fall into two categories: acellular matrices, where matrices are used alone and depend on the body's natural ability to regenerate for proper orientation and direction of new tissue growth, and matrices with cells. Acellular tissue matrices are usually prepared by removing cellular components from tissues via mechanical and chemical manipulation to produce collagen-rich matrices. These matrices tend to slowly degrade on implantation and are generally replaced by the ECM proteins that are secreted by the ingrowing cells. When cells are used for tissue engineering, a small piece of donor tissue is dissociated into individual cells.

These cells are either implanted directly into the host or are expanded in culture, attached to a support matrix, and then reimplanted into the host after expansion. The source of donor tissue can be heterologous (such as bovine), allogeneic (same species, different individual), or autologous. Ideally, both structural and functional tissue replacements will occur with minimal complications. The most preferred cells to use are autologous cells, where a biopsy of tissue is obtained from the host, the cells are dissociated and expanded in culture, and the expanded cells are implanted into the same host. The use of autologous cells avoids rejection, and thus the deleterious side effects of immunosuppressive medications can be avoided.

One of the limitations of applying cell-based regenerative medicine techniques toward organ replacement has been the inherent difficulty of growing specific cell types in large quantities. Even when some organs, such as the liver, have a high regenerative capacity *in vivo*, cell growth and expansion *in vitro* may be difficult. By studying the privileged sites for committed precursor cells in specific organs, as well as exploring the conditions that promote differentiation, one may be able to overcome the obstacles that could lead to cell expansion *in vitro*. For example, urothelial cells could be grown in the laboratory setting in the past, but only with limited expansion. Several protocols were developed over the last two decades that identified the undifferentiated cells, and kept them undifferentiated during their growth phase.

By use of these methods of cell culture, it is now possible to expand a urothelial strain from a single specimen that initially covers a surface area of 1 cm^2 to one covering a surface area of 4202 m^2 (the equivalent area of one football field) within 8 wk. These studies indicated that it should be possible to collect autologous bladder cells from human patients, expand them in culture, and return them to the human donor in sufficient quantities for reconstructive purposes. Major advances have been achieved within the last decade on the possible expansion of a variety of primary human cells, with specific techniques that make the use of autologous cells possible for clinical application. For cell-based tissue engineering, the expanded cells are seeded onto a scaffold synthesized with the appropriate biomaterial.

In tissue engineering, biomaterials replicate the biologic and mechanical function of the native ECM found in tissues in the body by serving as an artificial ECM. As a result, biomaterials provide a three-dimensional space for the cells to form into new tissues with appropriate structure and function, and also can allow for the delivery of cells and appropriate bioactive factors (*e.g.*, cell adhesion peptides, growth factors), to desired sites in the body. Because the majority of mammalian cell types are anchorage dependent and will die if no cell-adhesion substrate is available, biomaterials provide a cell-adhesion substrate that can deliver cells to specific sites in the body with high loading efficiency. Biomaterials can also provide mechanical support against *in vivo* forces such that the predefined three-dimensional structure is maintained during tissue development. Furthermore, bioactive signals, such as cell-adhesion peptides and growth factors, can be loaded along with cells to help regulate cellular function.

The ideal biomaterial should be biocompatible in that it is biodegradable and bioresorbable to support the replacement of normal tissue without inflammation. Incompatible materials are destined for an inflammatory or foreign-body response that eventually leads to rejection and/or necrosis. In addition, the degradation products, if produced, should be removed from the body via metabolic pathways at an adequate rate that keeps the concentration of these degradation products in the tissues at a tolerable level.

Furthermore, the biomaterial should provide an environment in which appropriate regulation of cell behavior (*e.g.*, adhesion, proliferation, migration, and differentiation) can occur such that functional tissue can form. Cell behavior in the newly formed tissue has been shown to be regulated by multiple interactions of the cells with their microenvironment, including interactions with cell-adhesion ligands and with soluble growth factors. In addition, biomaterials provide temporary mechanical support that allows the tissue to grow in three dimensions while the cells undergo spatial tissue reorganization. The properly chosen biomaterial should allow the engineered tissue to maintain sufficient mechanical integrity to support itself in early development, whereas in late development, the properly chosen biomaterial should have begun degradation such that it does not hinder further tissue growth.

BIOMATERIALS USED FOR ENGINEERING TISSUES

Generally, three classes of biomaterials have been used for engineering tissues: naturally derived materials (*e.g.*, collagen and alginate), acellular tissue matrices (*e.g.*, bladder submucosa and small intestinal submucosa), and synthetic polymers (*e.g.*, polyglycolic acid (PGA), polylactic acid (PLA), and poly(lactic-co-glycolic acid) (PLGA). These classes of biomaterials have been tested in respect to their biocompatibility. Naturally derived materials and acellular tissue matrices have the potential advantage of biologic recognition. However, synthetic polymers can be produced reproducibly on a large scale with controlled properties of their strength, degradation rate, and microstructure.

Collagen is the most abundant and ubiquitous structural protein in the body, and may be readily purified from both animal and human tissues with an enzyme treatment and salt/acid extraction. Collagen implants degrade through a sequential attack by lysosomal enzymes. The *in vivo* resorption rate can be regulated by controlling the density of the implant and the extent of intermolecular cross-linking. The lower the density, the greater the interstitial space and generally the larger the pores for cell infiltration, leading to a higher rate of implant degradation. Collagen contains cell-adhesion domain sequences (*e.g.*, RGD) that exhibit specific cellular interactions. This may assist to retain the phenotype and activity of many types of cells, including fibroblasts and chondrocytes.

Alginate, a polysaccharide isolated from seaweed, has been used as an injectable cell delivery vehicle and a cell immobilization matrix because of its gentle gelling properties in the presence of divalent ions such as calcium. Alginate is relatively biocompatible and approved by the US Food and Drug Administration (FDA) for human use as wound dressing material. Alginate is a family of copolymers of D-mannuronate and L-guluronate. The physical and mechanical properties of alginate gel are strongly correlated with the proportion and length of polyguluronate block in the alginate chains.

Acellular tissue matrices are collagen-rich matrices prepared by removing cellular components from tissues. The matrices are often prepared by mechanical and chemical manipulation of a segment of tissue. The matrices slowly degrade upon implantation, and are replaced and remodeled by ECM proteins synthesized and secreted by transplanted or ingrowing cells.

Polyesters of naturally occurring á-hydroxy acids, including PGA, PLA, and PLGA, are widely used in tissue engineering. These polymers have gained FDA approval for human use in a variety of applications, including sutures. The ester bonds in these polymers are hydrolytically labile, and these polymers degrade by nonenzymatic hydrolysis. The degradation products of PGA, PLA, and PLGA are nontoxic, natural metabolites and are eventually eliminated from the body in the form of carbon dioxide and water.

The degradation rate of these polymers can be tailored from several weeks to several years by altering crystallinity, initial molecular weight, and the copolymer ratio of lactic to glycolic acid. Because these polymers are thermoplastics, they can be easily formed into a three-dimensional scaffold with a desired microstructure, gross shape, and dimension by various techniques, including molding, extrusion, solvent casting, phase separation techniques, and gas foaming techniques. Many applications in tissue engineering often require a scaffold with high porosity and ratio of surface area to volume. Other biodegradable synthetic polymers, including poly(anhydrides) and poly(ortho-esters), can also be used to fabricate scaffolds for tissue engineering with controlled properties.

HISTORICAL PERSPECTIVE

An important component in the early development of tissue engineering was the parallel development of artificial biomaterials. In the mid-1960s, artificial skin for burn victims was being pursued as a symptomatic therapy, and later, synthetic fibers were being tried as artificial skin grafts for burn treatment. In the early 1970s, there were concerted efforts to treat artificial surfaces to be used in implants in ways that would enable them to avoid causing blood coagulation, by applying special heparin complex coatings, for example. Other efforts focused on the toxicology profiles and biocompatibility of a variety of organic polymers considered for implants or tissue engineering, and the development of novel gels as the basis for artificial skin. In the late 1970s, researchers experimented with collagen-based artificial skin for use in oral mucosa injuries.

In the 1980s, R&D in tissue engineering and biomaterials took off. As part of this interest, several biomedical engineering departments were established at major universities around the world. In 1981, a skin equivalent consisting of a silicone cover over a sponge of porous collagen cross-linked with chondroitin was used successfully to treat severe burns.

In this decade, several products reached the market. Interpore's Pro-Osteon coral-derived bone graft material was introduced in 1993. In 1996, Integra's Artificial Skin was approved for as an in vivo, nonbiological tissue regeneration product. Then, in 1998, the General and Plastic Surgery Devices Advisory Panel to the US Food and Drug Administration recommended unconditional approval of Apligraf (Graftskin) Human Skin Equivalent for the treatment of venous leg ulcers. Apligraf, produced by Organogenesis, is the first manufactured living human organ, specifically multilayered skin, to be recommended for approval by an advisory panel to the FDA. Apligraf was approved for the treatment of venous leg ulcers in Canada in 1997, and was launched there in August 1997 by Novartis Pharmaceuticals Canada (Dorval, Canada).

Current State

Medtronic, a leading medical device company with a cardiac surgery business, offers a complete line of mechanical and tissue prosthetic heart valves, and the alliance with LifeCell gives it access to a significant tissue-engineered product to complement its own lineup. This type of alliance illustrates an alternate route to the big pharma path for commercializing tissue engineering products—that is, through more specialized medical device companies. Part of the interest and support for tissue engineering comes from the armed forces, in that numerous battlefield-related medical applications exist for tissue-engineered products and biomaterials.

For example, Advanced Tissue Sciences had part of its clinical trial for Dermagraft-TC in the treatment of chemical burns funded by the US Army Institute of Chemical Defense. Dermagraft-TC is an engineered human dermal tissue combined with a synthetic epidermal layer. It covers and protects burns, helping to minimize infections and retain fluids until a sufficient amount of the patient's own skin is available for autologous grafting. The principal alternative is cadaver skin, but the problems here include a limited supply, acute immunological rejection, and potential pathogen transmission.

Another application of tissue-engineered products is in the toxicology testing and in vitro markets, as alternatives to certain types of animal testing. A good example of this approach was the acquisition in 1995 by Stratum Laboratories (La Jolla, CA) of the In Vitro Laboratory Testing (IVLT) business of Advanced Tissue Sciences. Stratum received license rights to manufacture and sell Skin2 in vitro laboratory testing kits, in addition to an option to extend rights for up to six additional tissues. Skin2 is living human skin tissue used to test skin care, household, chemical, and pharmaceutical products for a variety of indications. The material is already approved by the US Department of Transportation and Transport Canada for use in a corrosivity test, demonstrating significant regulatory acceptance of this technology.

The use of progenitor-type cells as the starting point for developing differentiated tissue material is the focus of considerable research. For example, it is possible to take mesenchymal stem cells that reside in the adult bone marrow and induce them to differentiate into chondrocytes by using specific tissue culture media that include transforming growth factor β. Chondrocytes are constituents of cartilagenous tissue, and the possibility of generating them in a controlled fashion creates possibilities for the development of appropriate cartilage tissue for surgical procedures. Stem cells are also used as a starting point for a multitude of other cell types used in tissue engineering. Wound repair is a key application for tissue engineering products. Although most applications focus on the use of artificial skin to treat burns, different disease conditions are also benefiting.

For example, Advanced Tissue Sciences' Dermagraft is a three-dimensional human neonatal dermal fibroblast culture that has been grown on a biodegradable scaffold and cryopreserved. It has been applied to foot ulcers that develop as a side effect of long-term diabetes. In clinical trials, significant healing occurred with this material, especially when the Dermagraft cells were alive and functioning properly. Finally, increasing effort is being focused on correlating the actual physics of engineered cells that are to be used therapeutically. One recent report describes how the pressure that is applied to chondrocytes transplanted into articular cartilage defects is likely to inhibit the growth of these cells. Thus, the actual mechanics of the environment into which bio-engineered cells are used needs to be carefully considered to ensure that optimal benefit is derived from these approaches.

Industry Challenges

Quality control of the materials used in various surgical applications is a key challenge for the tissue engineering industry. For example, living human cells are being used in scaffolds to repair structural tissue damage. These materials need to be produced and cultured under good manufacturing practice (GMP) conditions to meet FDA standards—especially the cells that are grown ex vivo. As a result, the tissue engineering industry is striving to create appropriate quality control standards and evaluate them. A good example is that of autologous cultured chondrocytes used to repair knee damage. Recently, Genzyme Tissue Repair reported on a quality control program for this material that was based on evaluating 303 patients who had received the material.

The program was based on the analysis of several quantifiable parameters, including viability and freedom from pathogens; results showed that the materials were, indeed, appropriate for their use, and demonstrated one way to follow up with quality control monitoring of tissue-engineered products. Another challenge concerns acquiring a fundamental understanding of tissue differentiation mechanisms that might be harnessed for the development of tissue-engineered products. One product that the tissue engineering industry is pursuing is "bone-on-demand," to be used in cases where new bone formation is needed.

An important component here is the bone morphogenetic protein (BMP) complex, which is capable of inducing extraskeletal bone formation at a concentration 1,000-fold lower than each of its constituents alone. Researchers are trying to concentrate BMP complex locally, using appropriate implants rather than the individual constituents, which could potentially lead to bone formation. Finally, the industry is challenged to develop tissue-engineered products for a number of surgery-related applications, such as vasculature.

For example, a recent report describes the generation of a tissue-engineered blood vessel without any synthetic or exogenous materials. The vessels were produced by wrapping a sheet of human vascular smooth muscle cells around a tubular support, followed by wrapping a sheet of human fibroblasts around the first sheet. After cell maturation, the tubular support was removed and endothelial cells were seeded in the lumen of the putative blood vessel. The vessel showed all the necessary key markers of activity, and also had a burst strength comparable to human vessels. In short-term grafts in a dog model, this engineered vessel demonstrated good handling and saturability characteristics.

The Future

Tissue engineering has significant market potential and financial investment continues apace. A 1997 survey of the field reported that in that year alone, R&D expenditure directly linked to corporate tissue engineering projects was about $0.5 billion, with a growth rate of about 22 per cent per year. This demonstrates the sustained interest in this area, driven in part by positive results regarding specific products and processes in clinical settings. Technical advances in the various components of the industry will contribute to market growth. One component is the availability of biomaterials that act as scaffolds for tissue repair and reconstruction, or for the deposition of engineered tissues and cells preceding implantation. An increasing amount of R&D is directed toward addressing the properties of these scaffolds with the goal of creating materials that have the desired functional profiles for various applications.

For example, so-called blended-polymer scaffolds have an extended lifetime in the body that is more suitable for orthopedic applications than nonblended scaffolds. Another study has shown how collagen-based scaffolds, used to grow keratinocytes in artificial skin preparations, can be manipulated by cross-linking collagen with glycosaminoglycan. The result was increased biological stability, which augmented the likelihood that the keratinocytes would "take" and grow out of the scaffold. In yet another application, a sacchitin glycolipid-type membrane prepared from the residue of a fungal fruiting body has been shown to have significant promise in animal models of skin damage as a skin substitute that facilitated wound healing and fibroblast growth.

Development of new materials of this type that enable different applications of tissue engineering is likely to be the focus of considerable future research. For the biological component of tissue engineering, rapid advances are being made in identifying new cell types for use in tissue regeneration.

For example, undifferentiated stem cells are attracting intense interest because of their capacity to be transformed into almost any cell type that may be needed, and even fat cells can be directed to produce appropriate tissues. In addition, promising artificial nerve grafts or nerve guidance channels are being developed for nerve regeneration.

In the future, efforts will likely increasingly focus on the development of tissue-engineered products under consensus safety and efficacy standards, including sourcing of cells and tissues, characterization and testing of the materials, quality assurance and control, and preclinical and clinical evaluation.

The FDA has already provided some regulatory guidance concerning specific materials, such as certain marketed artificial skin products; in the next few years, these guidelines will likely be increasingly formalized and structured, ensuring that tissue engineering products not only work but are also safe. Significant future developments will include the continued development of artificial organs that use cells embedded into appropriate support structures. A recent report describes the use of polyurethane foam as a good matrix into which liver cells grew as spheroids, the system showing promise as an artificial liver.

The future will also see significant efforts to develop engineered vascular grafts. An approach that will see increasing attention is that of taking a scaffold that is a structurally intact xenogeneic vessel, such as a pig aorta, removing all cells, and re-populating this with human autologous cells. A recent report showed how this could be done in 2–3 weeks, opening up the way for a good alternative to vascular engineering.

The range of human tissue that can be engineered will also increase dramatically in the future, so in addition to the traditional targets, such as skin and liver, other tissues and organs will see their day. A great deal of excitement in clinical circles is that of developing artificial human thyroid tissues which are capable of producing T cells, and this will be a major area for continued R&D.

Finally, stem cells and their manipulation for therapeutic purposes will continue to be a major area of development, because of the pluripotency of these cells. For example, bone marrow stem cells contained in resorbable artificial tubes have been shown to lead to effective healing of non-union defects in rabbit radii, and this opens up significant surgical alternatives to organ and tissue damage.

SCAFFOLDING APPROACHES IN TISSUE ENGINEERING

Over the last two decades, four major scaffolding approaches for tissue engineering have evolved. It would be informative to highlight the working principles and the characteristics of these approaches before discussing the tissue-specific considerations.

Pre-made Porous Scaffolds for Cell Seeding

Since the birth of "tissue engineering", seeding therapeutic cells in pre-made porous scaffolds made of degradable biomaterials has become the most commonly used and well-established scaffolding approach. This approach represents the bulk of biomaterial research in tissue engineering, leading to enormous efforts in development of different types of biomaterials and fabrication technologies.

Many types of biomaterials can be used to make porous scaffolds for tissue engineering provided that a fabrication technology compatible with the biomaterial properties is available. Reviewing these biomaterials in detail is out of the scope of this paper and readers are directed to other reviews. In general, biomaterials used for making porous scaffolds for tissue engineering can be classified into two categories according to their sources, namely natural and synthetic biomaterials. Naturally occurring biomaterials can be obtained from their natural sources and processed to make porous scaffolds.

These materials can be in their native form, such as ECM from allografts and xenografts, or can be in the form of smaller building blocks, which include but not limited to inorganic ceramics such as calcium phosphates and organic polymers such as proteins, polysaccharides, lipids and polynucleotides. Natural biomaterials usually have superb biocompatibility so that cells can attach and grow with excellent viability. However, one issue with natural materials is their limited physical and mechanical stability and therefore they may not be suitable for some load-bearing applications.

This is the reason why researchers using natural biomaterials are prompted to develop technologies improving and reinforcing the mechanical and shape stability of natural biomaterials. Examples include developing composites with synthetic material and crosslinking. Another issue is the potential immunogenicity, because natural biomaterials from allogenic or xenogenic sources may be antigenic to the hosts. As a result, researchers are attracted to technologies such as removal of telopeptides in procollagen for reduction of immunogenicity. Synthetic biomaterials can be generally categorized into inorganic such as bioglasses, and organic such as synthetic polymers.

It is generally believed that synthetic biomaterials have better controlled physical and mechanical properties and can be used to tailor for both soft and hard tissues. Nevertheless, for synthetic biomaterials, biocompatibility becomes the major issue because cells may have difficulties in attachment and growth on these materials. Therefore, many processes modifying the surface and bulk properties have been developed to improve their biocompatibility. Examples include surface laser engineering and coating with natural biomaterials such as collagen. With the development of composites materials, the combinations of biomaterials for making porous scaffolds have become enormous. Multiple types of biomaterials with distinct properties can be blended and fabricated to be tailored for a particular application.

Over the last decade, development of fabrication technologies for porous scaffolds has been an intensive area of research. Readers are also directed to several excellent review articles for various fabrication technologies.

In general, these technologies can be classified into:

- Processes using porogens in biomaterials,
- Solid free-form or rapid prototyping technologies and
- Techniques using woven or non-woven fibers.

In the first category, solid materials either in solids or dissolved in solvents, are incorporated with porogens, which could be gases such as carbon dioxide, liquids such as water or solids such as paraffin. The mixtures are processed and then cast or extruded. Porogens are removed after fabrication using methods such as sublimation, evaporation and melting to leave behind a porous structure in the scaffold. Example techniques include solvent casting and particulate leaching, gas foaming, freeze-drying and phase separation. In the second category, hierarchical porous structures are manufactured by sequential delivery of material and/or energy needed to bond the materials to preset points in space.

Some solid free-form fabrication technologies such as selective laser sintering, stereolithography and 3D printing depend on precise delivery of light or heat energy in a scanner system to points of space in the material bed so as to bond or crosslink the materials to give solid structures in an otherwise soluble bed of materials. Some solid free-form fabrication technologies such as wax printing rely on simultaneous delivery of solid materials and removable supporting materials in a layer-by-layer printing manner. In the third category, woven and non-woven fiber structures can be piled together and bonded using thermal energy or adhesives to give a porous meshwork using techniques such as fiber bonding, or fibers can be generated by the electrospinning technique, in which a high voltage is injected to a polymer solution where the electrostatic forces are built up to overcome the surface tension of the polymer solution and therefore form a spinning fiber jet.

This scaffolding approach using pre-made porous scaffolds has a number of advantages. Firstly, this approach has the most diversified choice for biomaterials. Almost all biomaterials ranged from ceramics to hydrogels have suitable fabrication technologies. Secondly, relatively precise designs on architecture and microstructure of the scaffolds can be incorporated by this approach. Therefore, the physicochemical properties of the porous scaffolds are easily engineered in order to mimic the physical properties of native ECM in the target tissues. This is especially advantageous for making load-bearing tissues where the mechanical properties are important. Nevertheless, this approach also has certain disadvantages. In particular, post-fabrication cell-seeding to porous scaffolds is time-consuming and inefficient because of the limited penetration ability of cells into the scaffolds. This usually leads to inhomogeneous distribution of cells in the scaffolds and therefore heterogeneous properties in the engineered tissues. As a result, various efforts such as agitation, perfusion and enlarged pore size are needed to enhance cell seeding efficiency and these methods are not free of problems, to list a few, low cell viability and high cost.

Decellularized ECM from Allogenic or Xenogenic Tissues for Cell Seeding

Acellular ECM processed from allogenic or xenogenic tissues are the most nature-simulating scaffolds, which have been used in tissue engineering of many tissues including heart valves, vessels, nerves, tendon and ligament. This scaffolding approach removes the allogenic or xenogenic cellular antigens from the tissues as they are the sources for immunogenicity upon implantation but preserves the ECM components, which are conserved among species and therefore well tolerated immunologically. Specialized decellularization techniques are developed to remove cellular components and this is usually achieved by a combination of physical, chemical and enzymatic methods.

In brief, cell membranes are lysed by physical treatments such as freeze-thaw cycles or ionic solutions such as hypo or hypertonic solutions before separating the cellular components from the ECM by enzymatic methods such as trypsin/EDTA treatment. The cytoplasmic and nuclear cellular components will then be solubilized and removed by a handful choice of detergents. The process parameters need to be optimized to aid complete decellularization with minimal disturbance on biochemical composition and mechanical properties of the ECM. Decellularized ECM can be used for homologous functions when the decellularized ECM is used to replace an analogous structural tissue that has been damaged.

An example is to decellularize vessels as allogenic vascular grafts. Decellularized ECM can also be used for non-homologous functions when it is used for a purpose different from which it fulfills in its native state, or in a location of the body where such structural function does not normally occur. Examples are to use small intestinal submucosa (SIS) for vascular graft, tendon, dura mater, skin and other tissues and to use amnion membrane for peripheral nerve regeneration. The major advantage of this scaffolding approach is the most close-to-nature mechanical and biological properties of the decellularized ECM. Moreover, apart from the excellent biocompatibility of the natural ECM, growth factors preserved in the decellularized matrix may further facilitate cell growth and remodeling. Nevertheless, cell seeding in decellularized ECM may also lead to inhomogeneous distribution while incomplete removal of cellular components may elicit immune reactions upon implantation.

Cell Sheets with Self-secreted ECM

Cell sheet engineering represents an approach where cells secrete their own ECM upon confluence and are harvested without the use of enzymatic methods. This is achieved by culturing cells on thermo-responsive polymer, such as poly(*N*-isopropylacrylamide) coated culture dish until confluence. The confluent cell sheet is then detached by thermally regulating the hydrophobicity of the polymer coatings without enzymatic treatment. Such approach can be repeated to laminate multiple single cell layers to form thicker matrix.

This technology is pioneered by Japanese group and has been systemically applied to a number of applications such as cornea in clinical trials and myocardium in preclinical trials. The same approach has been modified and improved to produce patterned substrates for thicker tissue fabrication and for injectable applications. Thermosensitive chitosan has also been investigated to create an aligned cell sheet. Cell sheet engineering approach is excellent for epithelium, endothelium and cell-dense tissues as the formation of cell sheets require cells to grow into confluence at high density such that cells can form tight junctions with each other and secrete ECM of their own. High cell density and close association is a characteristic of epithelium and endothelium.

Therefore, corneal epithelium and endothelium, vessel endothelium and tracheal epithelium are good candidates of this approach. Another advantage of the cell sheet engineering approach is that the laminated layers resulted in rapid neovascularization unlike transplantation of thick constructs with cell-seeded scaffolds. A recent report reviewed other advantages of the cell sheet engineering approach such as easy harvesting procedure, possibility of sutureless transplantation, and even distribution of cells in the sheet without mass transfer problem. Nevertheless, one disadvantage of this approach is that it is difficult to construct thick tissues as each layer is around 30 ìm thick. Multisurgeries such as 10 layers to form a 300 ìm thick myocardial patch are required.

This is less clinically feasible as multiple surgeries in patients are unlikely and therefore preformed patches are needed. The other disadvantage or perhaps limitation of cell sheet engineering approach is that constructing ECM rich tissues and hypocellular tissue such as bones, cartilage and intervertebral disc are unlikely as the amount of ECM secreted upon cell confluence is severely limited. In another word, tissues with rich ECM for load bearing purposes are unlikely to be fabricated by a cell sheet engineering approach.

Cell Encapsulation in Self-assembled Hydrogel Matrix

Encapsulation is a process entrapping living cells within the confines of a semi-permeable membrane or within a homogenous solid mass. The biomaterials used for encapsulation are usually hydrogels, which are formed by covalent or ionic crosslinking of water-soluble polymers. Many types of biomaterials including natural and synthetic hydrogels can be used for encapsulation provided that the conditions inducing the hydrogel formation or the polymerization are compatible with living cells. Encapsulation has been developed over several decades and the predominating use is for immunoisolation during allogenic or xenogenic cell transplantation.

Naturally occurring polysaccharides derived from algae, sodium alginate is the most commonly used material while other natural materials such as agarose and chitosan and synthetic materials such as poly (ethylene glycol) (PEG) and polyvinylalcohol (PVA) are also used. The most well-known application is xenogenic pancreatic cell transplantation for diabetes while applications for other disorders such as CNS insufficiency and liver failure are also reported. For immunoisolation to work, biomaterials encapsulating the cells need to be crosslinked or processed to become impenetrable to cells, impermeable to large molecules such as antibodies and cellular antigens but permeable to nutrients such as oxygen and glucose, metabolites such as carbon dioxide and lactic acid, and secreted therapeutic biomolecules from the encapsulated cells such as insulin from pancreatic beta cells.

In case of a semi-permeable membrane, the encapsulated cells should have the ability to maintain viability and functionality in aggregates even though there is only solid anchorage support limited to the luminal surface. Pancreatic cells, hepatocytes and haematopoietic cells are of this type. In case of a homogenous solid mass, where the entrapped cells are closely interacting with the biomaterials, the biomaterials should have good biocompatibility enabling cellular attachment and growth. Nevertheless, the commonly used encapsulating materials such as alginate and agarose have limited ability to support cell attachment growth and differentiation, resulting in low cell viability and growth. In many cases, a biomaterial with better biocompatibility such as collagen must be supplemented for improvement in cell viability.

Collagen is a natural biocompatible and biodegradable material and can be reconstituted into fibrous structures simulating the native ECM in tissues. Recently, a microencapsulation system immobilizing living cells within reconstituted collagen fiber meshwork has been established and the collagen meshwork is able to provide a bio-mimetic scaffold supporting cell growth, migration, therapeutic protein secretion and stem cell differentiation. Moreover, chemical approaches have been used to design self-assembled peptides and these biomimetic peptides can also be used to entrap cells. One important feature of encapsulation is that the biomaterials used are able to self-assemble from liquid monomers to solid polymer meshwork upon initiation, which is usually pH, temperature, ionic strength and light controlled.

To list a few examples, alginate solidifies when its monomer solution is exposed to divalent ion solutions such as calcium chloride, where the calcium ion crosslinks the alginate; collagen monomers polymerize when they are switched from an acidic pH and a low temperature to a neutral pH and a body temperature; ethylene glycol modified with acrylic moiety starts to polymerize when it is exposed to UV light in the presence of a photo-initiator. This unique feature combines the scaffold fabrication and the cell seeding into one-step procedure as cells can be mixed with the liquid biomaterials before initiation of polymerization. The advantages of this approach include simple one-step procedure, homogenous cell distribution in the hydrogel and excellent cell viability.

Moreover, this self-assembled approach enables injectable application where the polymerization can be initiated after injection, leading to "setting" of the hydrogel after they are injected into the defective tissues. This represents a minimally invasive approach of tissue engineering and is advantageous when the defect is irregularly shaped. Because of the injectable feature of the scaffold, this approach is sometimes referred as injectable scaffolding or in situ tissue engineering. Nevertheless, hydrogel materials in this approach used usually have poor mechanical properties. As a result, this scaffolding approach is seldom used for tissues with load bearing functions.

Scaffolding in Intervertebral Disc Tissue Engineering

Selecting the scaffolding approach for tissue engineering is tissue- and application-specific. Intervertebral disc is chosen. Extensive efforts have been made to search for biological therapeutics for disc degeneration of different severity. Readers are directed to excellent reviews on structural functional relationship and pathophysiological aspects of IVD and excellent reviews on potential biological therapies including growth factor, cell and tissue engineering approaches. In this review, existing scaffolding approaches for disc regeneration and their insufficiencies, the unique considerations of intervertebral disc and the future directions of scaffolding in IVD tissue engineering.

In early disc degeneration, disc cells in particular the nucleus pulposus (NP) cells become less capable to synthesize the proper ECM. The gelatinous NP becomes more fibrous with reduced water content. The primary to be repaired should be the capability of the NP cells to secrete the right matrix components in particular proteoglycans, which are responsible for the water absorbing function of the nucleus matrix. At this early stage, matrix loss is still limited or partial and there is no need to replace the bulk matrix. As a result, minimally invasive treatments such as injecting growth factors to stimulate the NP cells to synthesize proteoglycans or injecting cells able to synthesize appropriate ECM such as bone marrow mesenchymal stem cells are the most commonly proposed.

Nevertheless, one important inadequacy of this approach is that the high disc pressure and the aqueous nature of the cell suspension usually result in depletion of the local availability of cells. As a result, use of injectable carriers is necessary to effectively deliver cells into the degenerative NP space. Therefore, self-assembled hydrogels able to suspend and deliver cells in solution via injection but which can solidify after injection presents an appropriate scaffolding approach. An example is to deliver mesenchymal stem cells in hyaluronan gel. Nevertheless, these hydrogel carriers still have insufficient viscosity and stiffness resulting in immediate loss of the majority of injected cells (>96 per cent) due to back-flow via the injection path. This has been observed in our own group using other hydrogels (unpublished work). Therefore, better injectable approaches such as carriers with higher viscosity and stiffness, and in situ welding techniques such as laser welding, photochemical welding and use of biological glues at the injection site should be explored in order to prevent leakage and extrusion and thus improve the effectiveness of the injectable therapy.

As the disc degeneration progresses, more ECM and structural changes such as proteoglycan and collagen degradation involving NP and sometimes annulus fibrosus (AF) ensue, leading to disc height reduction. At this stage, replacement of cells with capacity to synthesize ECM and compensation of the substantial loss of matrix components are appropriate therapeutic approaches. This can be achieved by implanting cell-seeded pre-made scaffolds (approach 1) or decellularized ECM (approach 2) or injecting cells encapsulated in self-assembled hydrogels (approach 4) into the intradiscal space. In reviewing the existing scaffolding approaches for IVD tissue engineering, approach 1 using pre-made scaffold is most common.

A wide range of biomaterials dominated by natural biomaterials has been used for nucleus replacement. These materials include collagen, atellocollagen, alginate, gelatin, chitosan, collagen/glycosaminoglycan, collagen/hyaluronan and poly-L-lactic acid. Reviews for nucleus replacements have been reported elsewhere. On the other hand, biomaterials used for annulus replacement are dominated by synthetic biomaterials including silk, polycaprolactone and its derivatives, polyglycolic acid/polylactic acid and bioglass with a few natural biomaterials such as collagen/hyaluronan. In most if not all cases, freeze-drying has been employed as the fabrication method to create porous structures.

In most nucleus and annulus replacement strategies, survival and growth of seeded cells, and enhanced synthesis of collagen II and proteoglycans by these cells are reported. Nevertheless, almost all studies using the pre-made scaffold approach are in vitro while only a few are in vivo. Subcutaneous implantation was used and the implant was never exposed to physiological loading. Recently, there is one ex vivo study reporting collagen nucleus replacement in bovine discs under mechanical loading. Although restoration of disc height is possible, extrusion of the whole implant after a few loading cycles have been reported.

As a result, maintaining annulus integrity by sealing or welding methods after nucleotomy or injection or insertion of nucleus replacements is crucial. As for annulus replacement, the intrinsic mechanical properties of the annulus replacements have to be comparable to that of the native disc in order to be able to resist physiological loading. This is extremely challenging because the native annulus fibrosis serves a highly demanding and complex mechanical function. In existing literature, only a few annulus replacement strategies mentioned enhanced mechanical properties using composites materials and electrospun fibers with alignment. As a result, further efforts in enhancing the mechanical integrity of scaffolds for annulus replacement should be encouraged.

The efforts using the second scaffolding approach, decellularized ECM for nucleus replacement is minimal. Only one report uses porcine SIS as the scaffold for human disc cells. Cell survival and enhanced matrix deposition have been reported in vitro. In a pilot animal study using SIS in nucleotomized baboons, MRI suggested a higher water content in the treatment groups compared to the nucleotomy group and some tissue remodeling in the disc space in the group with bone marrow-soaked SIS. Moreover, SIS has good tissue biocompatibility in the disc space and seemed to have biodegraded over the six-month period. Nevertheless, a sizable study with more animals is needed before a conclusive statement on the efficacy of using SIS as nucleus replacement can be made.

There is no study in decellularizing allogenic or xenogenic nucleus or annulus for replacement at all. Nevertheless, there is one study evaluating the mechanical properties of decellularized temporomandibular joint disc. Although the immune privileged status of IVD, which is avascular, allows the application of allografts without decellularization, research efforts in decellularizing nucleus and annulus grafts from xenogenic sources for scaffolding should be encouraged because allografts are not always available. Preservation of proteoglycans will, however, be challenging. Recently, there is a surge in number of papers using injectable nucleus replacement, which is based on the fourth scaffolding approach, where cells are encapsulated in self-assembled biomaterials and injected to the disc space before the gel "sets". Mesenchymal stem cells and disc cells have been encapsulated in thermosensitive hydroxybutyl chitosan gel and cell proliferation and matrix production has been demonstrated.

Atellocollagen type II, hyaluronan and aggrecan gels support NP viability. Another study using atellocollagen shows that atellocollagen is better than alginate in supporting NP cells with better water and proteoglycan retention. Nevertheless, most studies on injectable modality using approach 4 are in vitro. Encouraging in vivo studies have been relatively scarce. In a rabbit model, MSCs were loaded in solution atellocollagen and injected to degenerative discs. Disc degeneration is effectively arrested by the treatment, and evidence of atellocollagen supporting cell growth, differentiation and matrix production is demonstrated. In a pig model, MSCs were loaded with hyaluronan derivatives and injected into the nucleotomized discs.

Close similarity in disc biconvex structure and viable chondrocytes like cells were reported. Further enhancement in the swelling and mechanical stability of the scaffolds for nucleus or annulus replacement such as crosslinking without compromising cell viability should be encouraged as these physical properties favor the restoration and maintenance of disc height. Evaluation of the mechanical properties of injectable polymers such as hyaluronic acid gel and polyethylene glycol against different engineering parameters should also be encouraged as replacing the mechanical function is equally important as replacing the cellular function in matrix secretion in this stage of degeneration.

At the advanced stage of disc degeneration, structural collapse and eventual loss of disc function in resisting loading set in motion. At this stage, probably only an engineered IVD tissue idealized in the discussion above can do the job. Decellularized IVD allograft is theoretically the best scaffold for late stage replacement. To date, there is no decellularization study in intervertebral disc, but there are several in vivo studies transplanting undecellularized allogenic discs. A few groups have also transplanted cryopreserved allogenic disc grafts in dogs. Nevertheless, dramatic decrease in cellular activity has been reported, indicating that preservation of cellular synthetic and remodeling activities is important for long term viability of the grafts.

Furthermore, degenerative changes have been reported at 1 year post-implantation, also suggesting the need to preserve better cellular remodeling capacity. Larger animal model in monkeys showed that fresh frozen allografts can maintain the mechanical properties and some degree of cell metabolism, but severe degeneration has been observed in 2 years time, associating with decreased biochemical contents of the disc. This study also suggests the need to preserve or supplement viable cells with remodeling capacity that is able to maintain long term merits of allografting. Very recently, this group conducted a first allograft disc segment transplantation study in human. The transplanted segments were able to preserve motion and hydration for at least 5 years with partially recovered disc height and improved neurological symptoms, and with no immunoreaction.

Nevertheless, mild degenerative changes have been found after years of follow-up, suggesting again that repopulation or supplementation of live cells with matrix remodeling capacity is necessary for long term functionality of allografts. As a result, a promising direction is to develop better graft preservation technologies for maintaining the disc cell remodeling capacity. Alternatively, supplementing stem cells or other cells with the ability to respond to physiological challenges in terms of synthesis and secretion of appropriate ECM into the allograft disc space is also promising. In this regard, searching for injectable carriers to provide an appropriate microenvironment to the encapsulated cells for proper remodeling response toward mechanical stimulation deserves more attention.

Apart from using the allograft, building an implantable IVD with multiple tissue components with mechanical properties comparable with the native disc presents another possibility. Activities with this line of research have been scarce, and only a few groups are attempting to build multiple tissue components and to target the interface problems. A few unique considerations should be highlighted as they are the most challenging tasks for tissue engineering. Firstly, IVD is a complex tissue with multiple tissue components.

It is unlikely to utilize single biomaterial and single scaffolding approach. Maintaining the differential hydration and mechanical properties of the nucleus and the annulus is important to the maintenance of the disc height and its load-resisting properties. Engineering designs to better retain the water absorbing capacity in the biomaterials used for nucleus replacement and engineering designs to better maintain the enclosure of the swelling nucleus within the confines of the annulus replacement warrant further attention. Secondly, IVD involves multiple tissue interfaces, which are essential for load transfer and distribution between hard and soft tissues and are crucial for maintaining proper disc functions.

Chondrocytes seeded between a pre-made bone scaffold and NP cells resulted in the formation of a cartilage-like layer between bone construct and nucleus cells while chondrocytes supplemented with osteogenic differentiation signal led to the formation of a calcified zone between bone and cartilage interface. Both studies demonstrated benefits in mechanical performance, in particular improving the interfacial shear stress. This suggests that engineering the interfaces among different tissue components should deserve special attention and more enthusiastic research efforts.

Disc tissue engineering is by default multidisciplinary, and the scaffolding approach is just one of the disciplines involved. Its success obviously relies on the corroborative efforts from other aspects. For examples, delineation of the etiology of disc degeneration, understanding of the disc nutrition mechanism, cell sourcing, identification of disc specific markers, growth-stimulating and differentiation-stimulating signals, etc. are also important aspects in achieving better functional outcomes of IVD tissue engineering.

STEM CELLS

Most current strategies for tissue engineering depend upon a sample of autologous cells from the diseased organ of the host. However, for many patients with extensive end-stage organ failure, a tissue biopsy may not yield enough normal cells for expansion and transplantation. In other instances, primary autologous human cells can not be expanded from a particular organ, such as the pancreas. In these situations, pluripotent human embryonic stem cells are envisioned as a viable source of cells because they can serve as an alternative source of cells from which the desired tissue can be derived.

Combining the techniques learned in tissue engineering over the past few decades with this potentially endless source of versatile cells could lead to novel sources of replacement organs. Embryonic stem cells exhibit two remarkable properties: the ability to proliferate in an undifferentiated but pluripotent state (self-renew), and the ability to differentiate into many specialized cell types. They can be isolated by immunosurgery from the inner cell mass of the embryo during the blastocyst stage (5 d after fertilization), and are usually grown on feeder layers consisting of mouse embryonic fibroblasts or human feeder cells. More recent reports have shown that these cells can be grown without the use of a feeder layer, and thus avoid the exposure of these human cells to mouse viruses and proteins. These cells have demonstrated longevity in culture by maintaining their undifferentiated state for at least 80 passages when grown using current published protocols.

Human embryonic stem cells have been shown to differentiate into cells from all three embryonic germ layers *in vitro*. Skin and neurons have been formed, indicating ectodermal differentiation. Blood, cardiac cells, cartilage, endothelial cells, and muscle have been formed, indicating mesodermal differentiation. And pancreatic cells have been formed, indicating endodermal differentiation. In addition, as further evidence of their pluripotency, embryonic stem cells can form embryoid bodies, which are cell aggregations that contain all three embryonic germ layers, while in culture, and can form teratomas *in vivo* .

WHAT ARE STEM CELLS, AND WHY ARE THEY IMPORTANT

Stem cells have the remarkable potential to develop into many different cell types in the body during early life and growth. In addition, in many tissues they serve as a sort of internal repair system, dividing essentially without limit to replenish other cells as long as the person or animal is still alive. When a stem cell divides, each new cell has the potential either to remain a stem cell or become another type of cell with a more specialized function, such as a muscle cell, a red blood cell, or a brain cell.

Stem cells are distinguished from other cell types by two important characteristics. First, they are unspecialized cells capable of renewing themselves through cell division, sometimes after long periods of inactivity. Second, under certain physiologic or experimental conditions, they can be induced to become tissue- or organ-specific cells with special functions. In some organs, such as the gut and bone marrow, stem cells regularly divide to repair and replace worn out or damaged tissues. In other organs, however, such as the pancreas and the heart, stem cells only divide under special conditions.

Until recently, scientists primarily worked with two kinds of stem cells from animals and humans: embryonic stem cells and non-embryonic "somatic" or "adult" stem cells. The functions and characteristics of these cells will be explained in this document. Scientists discovered ways to derive embryonic stem cells from early mouse embryos nearly 30 years ago, in 1981. The detailed study of the biology of mouse stem cells led to the discovery, in 1998, of a method to derive stem cells from human embryos and grow the cells in the laboratory. These cells are called human embryonic stem cells. The embryos used in these studies were created for reproductive purposes through *in vitro* fertilization procedures. When they were no longer needed for that purpose, they were donated for research with the informed consent of the donor. In 2006, researchers made another breakthrough by identifying conditions that would allow some specialized adult cells to be "reprogrammed" genetically to assume a stem cell-like state. This new type of stem cell, called induced pluripotent stem cells (iPSCs).

Stem cells are important for living organisms for many reasons. In the 3- to 5-day-old embryo, called a blastocyst, the inner cells give rise to the entire body of the organism, including all of the many specialized cell types and organs such as the heart, lung, skin, sperm, eggs and other tissues. In some adult tissues, such as bone marrow, muscle, and brain, discrete populations of adult stem cells generate replacements for cells that are lost through normal wear and tear, injury, or disease.

Given their unique regenerative abilities, stem cells offer new potentials for treating diseases such as diabetes, and heart disease. However, much work remains to be done in the laboratory and the clinic to understand how to use these cells for cell-based therapies to treat disease, which is also referred to as regenerative or reparative medicine. Laboratory studies of stem cells enable scientists to learn about the cells' essential properties and what makes them different from specialized cell types. Scientists are already using stem cells in the laboratory to screen new drugs and to develop model systems to study normal growth and identify the causes of birth defects.

Research on stem cells continues to advance knowledge about how an organism develops from a single cell and how healthy cells replace damaged cells in adult organisms. Stem cell research is one of the most fascinating areas of contemporary biology, but, as with many expanding fields of scientific inquiry, research on stem cells raises scientific questions as rapidly as it generates new discoveries.

UNIQUE PROPERTIES OF ALL STEM CELLS

Stem cells differ from other kinds of cells in the body. All stem cells—regardless of their source—have three general properties: they are capable of dividing and renewing themselves for long periods; they are unspecialized; and they can give rise to specialized cell types.

Stem cells are capable of dividing and renewing themselves for long periods. Unlike muscle cells, blood cells, or nerve cells—which do not normally replicate themselves—stem cells may replicate many times, or proliferate. A starting population of stem cells that proliferates for many months in the laboratory can yield millions of cells. If the resulting cells continue to be unspecialized, like the parent stem cells, the cells are said to be capable of long-term self-renewal.

Scientists are trying to understand two fundamental properties of stem cells that relate to their long-term self-renewal:

1. why can embryonic stem cells proliferate for a year or more in the laboratory without differentiating, but most non-embryonic stem cells cannot; and
2. what are the factors in living organisms that normally regulate stem cellproliferation and self-renewal?

Discovering the answers to these questions may make it possible to understand how cell proliferation is regulated during normal embryonic development or during the abnormal cell division that leads to cancer. Such information would also enable scientists to grow embryonic and non-embryonic stem cells more efficiently in the laboratory.

The specific factors and conditions that allow stem cells to remain unspecialized are of great interest to scientists. It has taken scientists many years of trial and error to learn to derive and maintain stem cells in the laboratory without them spontaneously differentiating into specific cell types. For example, it took two decades to learn how to grow human embryonic stem cells in the laboratory following the development of conditions for growing mouse stem cells. Therefore, understanding the signals in a mature organism that cause a stem cell population to proliferate and remain unspecialized until the cells are needed. Such information is critical for scientists to be able to grow large numbers of unspecialized stem cells in the laboratory for further experimentation.

Stem cells are unspecialized. One of the fundamental properties of a stem cell is that it does not have any tissue-specific structures that allow it to perform specialized functions. For example, a stem cell cannot work with its neighbors to pump blood through the body (like a heart muscle cell), and it cannot carry oxygen molecules through the bloodstream (like a red blood cell). However, unspecialized stem cells can give rise to specialized cells, including heart muscle cells, blood cells, or nerve cells.

Stem cells can give rise to specialized cells. When unspecialized stem cells give rise to specialized cells, the process is called differentiation. While differentiating, the cell usually goes through several stages, becoming more specialized at each step. Scientists are just beginning to understand the signals inside and outside cells that trigger each stem of the differentiation process. The internal signals are controlled by a cell's genes, which are interspersed across long strands of DNA, and carry coded instructions for all cellular structures and functions.

The external signals for cell differentiation include chemicals secreted by other cells, physical contact with neighboring cells, and certain molecules in the microenvironment. The interaction of signals during differentiation causes the cell's DNA to acquire epigenetic marks that restrict DNA expression in the cell and can be passed on through cell division.

Many questions about stem cell differentiation remain. For example, are the internal and external signals for cell differentiation similar for all kinds of stem cells? Can specific sets of signals be identified that promote differentiation into specific cell types? Addressing these questions may lead scientists to find new ways to control stem cell differentiation in the laboratory, thereby growing cells or tissues that can be used for specific purposes such as cell-based therapies or drug screening.

Adult stem cells typically generate the cell types of the tissue in which they reside. For example, a blood-forming adult stem cell in the bone marrow normally gives rise to the many types of blood cells. It is generally accepted that a blood-forming cell in the bone marrow—which is called a hematopoietic stem cell—cannot give rise to the cells of a very different tissue, such as nerve cells in the brain. Experiments over the last several years have purported to show that stem cells from one tissue may give rise to cell types of a completely different tissue. This remains an area of great debate within the research community. This controversy demonstrates the challenges of studying adult stem cells and suggests that additional research using adult stem cells is necessary to understand their full potential as future therapies.

EMBRYONIC STEM CELLS

Stages of Early Embryonic Development are Important for Generating Embryonic Stem Cells

Embryonic stem cells, as their name suggests, are derived from embryos. Most embryonic stem cells are derived from embryos that develop from eggs that have been fertilized *in vitro*—in an *in vitro* fertilization clinic—and then donated for research purposes with informed consent of the donors. They are *not* derived from eggs fertilized in a woman's body.

Embryonic Stem Cells Grown in the Laboratory

Growing cells in the laboratory is known as cell culture. Human embryonic stem cells (hESCs) are generated by transferring cells from a preimplantation-stage embryo into a plastic laboratory culture dish that contains a nutrient broth known as culture medium. The cells divide and spread over the surface of the dish. The inner surface of the culture dish is typically coated with mouse embryonic skin cells that have been treated so they will not divide. This coating layer of cells is called a feeder layer. The mouse cells in the bottom of the culture dish provide the cells a sticky surface to which they can attach. Also, the feeder cells release nutrients into the culture medium. Researchers have devised ways to grow embryonic stem cells without mouse feeder cells. This is a significant scientific advance because of the risk that viruses or other macromolecules in the mouse cells may be transmitted to the human cells.

The process of generating an embryonic stem cell line is somewhat inefficient, so lines are not produced each time cells from the preimplantation-stage embryo are placed into a culture dish. However, if the plated cells survive, divide and multiply enough to crowd the dish, they are removed gently and plated into several fresh culture dishes. The process of re-plating or subculturing the cells is repeated many times and for many months. Each cycle of subculturing the cells is referred to as a passage. Once the cell line is established, the original cells yield millions of embryonic stem cells. Embryonic stem cells that have proliferated in cell culture for for a prolonged period of time without differentiating, are pluripotent, and have not developed genetic abnormalities are referred to as an embryonic stem cell line. At any stage in the process, batches of cells can be frozen and shipped to other laboratories for further culture and experimentation.

Laboratory Tests are Used to Identify Embryonic Stem Cells

At various points during the process of generating embryonic stem cell lines, scientists test the cells to see whether they exhibit the fundamental properties that make them embryonic stem cells. This process is called characterization.

Scientists who study human embryonic stem cells have not yet agreed on a standard battery of tests that measure the cells' fundamental properties. However, laboratories that grow human embryonic stem cell lines use several kinds of tests, including:

- Growing and subculturing the stem cells for many months. This ensures that the cells are capable of long-term growth and self-renewal. Scientists inspect the cultures through a microscope to see that the cells look healthy and remainundifferentiated.
- Using specific techniques to determine the presence of transcription factors that are typically produced by undifferentiated cells. Two of the most important transcription factors are Nanog and Oct4. Transcription factors help turn geneson and off at the right time, which is an important part of the processes of celldifferentiation and embryonic development. In this case, both Oct 4 and Nanog are associated with maintaining the stem cells in an undifferentiated state, capable of self-renewal.
- Using specific techniques to determine the presence of paricular cell surface markers that are typically produced by undifferentiated cells.
- Examining the chromosomes under a microscope. This is a method to assess whether the chromosomes are damaged or if the number of chromosomes has changed. It does not detect genetic mutations in the cells.
- Determining whether the cells can be re-grown, or subcultured, after freezing, thawing, and re-plating.

- Testing whether the human embryonic stem cells are pluripotent by
  - Allowing the cells to differentiate spontaneously in cell culture;
  - Manipulating the cells so they will differentiate to form cells characteristic of the three germ layers; or
  - Injecting the cells into a mouse with a suppressed immune system to test for the formation of a benign tumor called a teratoma.

  Since the mouse's immune system is suppressed, the injected human stem cells are not rejected by the mouse immune system and scientists can observe growth and differentiation of the human stem cells. Teratomas typically contain a mixture of many differentiated or partly differentiated cell types—an indication that the embryonic stem cells are capable of differentiating into multiple cell types.

Embryonic Stem Cells Stimulated to Differentiate

As long as the embryonic stem cells in culture are grown under appropriate conditions, they can remain undifferentiated (unspecialized). But if cells are allowed to clump together to form embryoid bodies, they begin to differentiate spontaneously. They can form muscle cells, nerve cells, and many other cell types. Although spontaneous differentiation is a good indication that a culture of embryonic stem cells is healthy, it is not an efficient way to produce cultures of specific cell types.

So, to generate cultures of specific types of differentiated cells—heart muscle cells, blood cells, or nerve cells, for example—scientists try to control the differentiation of embryonic stem cells. They change the chemical composition of the culture medium, alter the surface of the culture dish, or modify the cells by inserting specific genes. Through years of experimentation, scientists have established some basic protocols or "recipes" for the directed differentiation of embryonic stem cells into some specific cell types.

If scientists can reliably direct the differentiation of embryonic stem cells into specific cell types, they may be able to use the resulting, differentiated cells to treat certain diseases in the future. Diseases that might be treated by transplanting cells generated from human embryonic stem cells include Parkinson's disease, diabetes, traumatic spinal cord injury, Duchenne's muscular dystrophy, heart disease, and vision and hearing loss.

ADULT STEM CELLS

An adult stem cell is thought to be an undifferentiated cell, found among differentiated cells in a tissue or organ that can renew itself and can differentiate to yield some or all of the major specialized cell types of the tissue or organ. The primary roles of adult stem cells in a living organism are to maintain and repair the tissue in which they are found. Scientists also use the term somatic stem cell instead of adult stem cell, where somatic refers to cells of the body (not the germ cells, sperm or eggs). Unlike embryonic stem cells, which are defined by their origin (cells from the preimplantation-stage embryo), the origin of adult stem cells in some mature tissues is still under investigation.

Research on adult stem cells has generated a great deal of excitement. Scientists have found adult stem cells in many more tissues than they once thought possible. This finding has led researchers and clinicians to ask whether adult stem cells could be used for transplants. In fact, adult hematopoietic, or blood-forming, stem cells from bone marrow have been used in transplants for 40 years. Scientists now have evidence that stem cells exist in the brain and the heart. If the differentiation of adult stem cells can be controlled in the laboratory, these cells may become the basis of transplantation-based therapies.

The history of research on adult stem cells began about 50 years ago. In the 1950s, researchers discovered that the bone marrow contains at least two kinds of stem cells. One population, called hematopoietic stem cells, forms all the types of blood cells in the body. A second population, called bone marrow stromal stem cells (also calledmesenchymal stem cells, or skeletal stem cells by some), were discovered a few years later. These non-hematopoietic stem cells make up a small proportion of the stromal cell population in the bone marrow, and can generate bone, cartilage, fat, cells that support the formation of blood, and fibrous connective tissue.

In the 1960s, scientists who were studying rats discovered two regions of the brain that contained dividing cells that ultimately become nerve cells. Despite these reports, most scientists believed that the adult brain could not generate new nerve cells. It was not until the 1990s that scientists agreed that the adult brain does contain stem cells that are able to generate the brain's three major cell types—astrocytes andoligodendrocytes, which are non-neuronal cells, and neurons, or nerve cells.

Adult Stem Cells found, and what do they Normally do

Adult stem cells have been identified in many organs and tissues, including brain, bone marrow, peripheral blood, blood vessels, skeletal muscle, skin, teeth, heart, gut, liver, ovarian epithelium, and testis. They are thought to reside in a specific area of each tissue (called a "stem cell niche"). In many tissues, current evidence suggests that some types of stem cells are pericytes, cells that compose the outermost layer of small blood vessels.

Stem cells may remain quiescent (non-dividing) for long periods of time until they are activated by a normal need for more cells to maintain tissues, or by disease or tissue injury. Typically, there is a very small number of stem cells in each tissue, and once removed from the body, their capacity to divide is limited, making generation of large quantities of stem cells difficult.

Scientists in many laboratories are trying to find better ways to grow large quantities of adult stem cells in cell culture and to manipulate them to generate specific cell types so they can be used to treat injury or disease. Some examples of potential treatments include regenerating bone using cells derived from bone marrow stroma, developing insulin-producing cells for type 1 diabetes, and repairing damaged heart muscle following a heart attack with cardiac muscle cells.

Tests are Used for Identifying Adult Stem Cells

Scientists often use one or more of the following methods to identify adult stem cells:

- Label the cells in a living tissue with molecular markers and then determine the specialized cell types they generate;
- Remove the cells from a living animal, label them in cell culture, and transplant them back into another animal to determine whether the cells replace (or "repopulate") their tissue of origin.

Importantly, it must be demonstrated that a single adult stem cell can generate a line of genetically identical cells that then gives rise to all the appropriate differentiated cell types of the tissue. To confirm experimentally that a putative adult stem cell is indeed a stem cell, scientists tend to show either that the cell can give rise to these genetically identical cells in culture, and/or that a purified population of these candidate stem cells can repopulate or reform the tissue after transplant into an animal.

Known about Adult Stem Cell Differentiation

As indicated above, scientists have reported that adult stem cells occur in many tissues and that they enter normal differentiation pathways to form the specialized cell types of the tissue in which they reside.

Normal differentiation pathways of adult stem cells. In a living animal, adult stem cells are available to divide, when needed, and can give rise to mature cell types that have characteristic shapes and specialized structures and functions of a particular tissue. The following are examples of differentiation pathways of adult stem cells that have been demonstrated *in vitro* or *in vivo*.

- Hematopoietic stem cells give rise to all the types of blood cells: red blood cells, B lymphocytes, T lymphocytes, natural killer cells, neutrophils, basophils, eosinophils, monocytes, and macrophages.
- Mesenchymal stem cells give rise to a variety of cell types: bone cells (osteocytes), cartilage cells (chondrocytes), fat cells (adipocytes), and other kinds of connective tissue cells such as those in tendons.
- Neural stem cells in the brain give rise to its three major cell types: nerve cells (neurons) and two categories of non-neuronal cells—astrocytes andoligodendrocytes.

- Epithelial stem cells in the lining of the digestive tract occur in deep crypts and give rise to several cell types: absorptive cells, goblet cells, paneth cells, and enteroendocrine cells.
- Skin stem cells occur in the basal layer of the epidermis and at the base of hair follicles. The epidermal stem cells give rise to keratinocytes, which migrate to the surface of the skin and form a protective layer. The follicular stem cells can give rise to both the hair follicle and to the epidermis.

Transdifferentiation: A number of experiments have reported that certain adult stem cell types can differentiate into cell types seen in organs or tissues other than those expected from the cells' predicted lineage (i.e., brain stem cells that differentiate into blood cells or blood-forming cells that differentiate into cardiac muscle cells, and so forth). This reported phenomenon is called transdifferentiation.

Although isolated instances of transdifferentiation have been observed in some vertebrate species, whether this phenomenon actually occurs in humans is under debate by the scientific community. Instead of transdifferentiation, the observed instances may involve fusion of a donor cell with a recipient cell. Another possibility is that transplanted stem cells are secreting factors that encourage the recipient's own stem cells to begin the repair process. Even when transdifferentiation has been detected, only a very small percentage of cells undergo the process.

In a variation of transdifferentiation experiments, scientists have recently demonstrated that certain adult cell types can be "reprogrammed" into other cell types in vivo using a well-controlled process of genetic modification. This strategy may offer a way to reprogram available cells into other cell types that have been lost or damaged due to disease. For example, one recent experiment shows how pancreatic beta cells, the insulin-producing cells that are lost or damaged in diabetes, could possibly be created by reprogramming other pancreatic cells.

By "re-starting" expression of three critical beta-cell genes in differentiated adult pancreatic exocrine cells, researchers were able to create beta cell-like cells that can secrete insulin. The reprogrammed cells were similar to beta cells in appearance, size, and shape; expressed genes characteristic of beta cells; and were able to partially restore blood sugar regulation in mice whose own beta cells had been chemically destroyed. While not transdifferentiation by definition, this method for reprogramming adult cells may be used as a model for directly reprogramming other adult cell types.

In addition to reprogramming cells to become a specific cell type, it is now possible to reprogram adult somatic cells to become like embryonic stem cells (induced pluripotent stem cells, iPSCs) through the introduction of embryonic genes. Thus, a source of cells can be generated that are specific to the donor, thereby increasing the chance of compatibility if such cells were to be used for tissue regeneration. However, like embryonic stem cells, determination of the methods by which iPSCs can be completely and reproducibly committed to appropriate cell lineages is still under investigation.

Key Questions about Adult Stem Cells

Many important questions about adult stem cells remain to be answered. *They include*:

- How many kinds of adult stem cells exist, and in which tissues do they exist?
- How do adult stem cells evolve during development and how are they maintained in the adult? Are they "leftover" embryonic stem cells, or do they arise in some other way?
- Why do stem cells remain in an undifferentiated state when all the cells around them have differentiated? What are the characteristics of their "niche" that controls their behavior?
- Do adult stem cells have the capacity to transdifferentiate, and is it possible to control this process to improve its reliability and efficiency?
- If the beneficial effect of adult stem cell transplantation is a trophic effect, what are the mechanisms? Is donor cell-recipient cell contact required, secretion of factors by the donor cell, or both?
- What are the factors that control adult stem cell proliferation and differentiation?
- What are the factors that stimulate stem cells to relocate to sites of injury or damage, and how can this process be enhanced for better healing?

CLONING

In biology, cloning is the process of producing similar populations of genetically identical individuals that occurs in nature when organisms such as bacteria, insects or plants reproduce asexually. Cloning in biotechnology refers to processes used to create copies of DNA fragments (molecular cloning), cells (cell cloning), or organisms. The term also refers to the production of multiple copies of a product such as digital media or software.

The term *clone* is derived from the Ancient Greek "twig", referring to the process whereby a new plant can be created from a twig. In horticulture, the spelling *clon* was used until the twentieth century; the final *e* came into use to indicate the vowel is a "long o" instead of a "short o". Since the term entered the popular lexicon in a more general context, the spelling*clone* has been used exclusively. In botany, the term lusus was traditionally used. In the United States, the human consumption of meat and other products from cloned animals was approved by the FDA on December 28, 2006, with no special labeling required because food from cloned organisms has been found to be identical to the organisms from which they were cloned. Such practice has met strong resistance in other regions due to misinformation, such as Europe, particularly over the labeling issue.

MOLECULAR CLONING

Molecular cloning refers to the process of making multiple molecules. Cloning is commonly used to amplify DNA fragments containing whole genes, but it can also be used to amplify any DNA sequence such as promoters, non-coding sequences and randomly fragmented DNA. It is used in a wide array of biological experiments and practical applications ranging from genetic fingerprinting to large scale protein production. Occasionally, the term cloning is misleadingly used to refer to the identification of thechromosomal location of a gene associated with a particular phenotype of interest, such as in positional cloning.

In practice, localization of the gene to a chromosome or genomic region does not necessarily enable one to isolate or amplify the relevant genomic sequence. To amplify any DNA sequence in a living organism, that sequence must be linked to an origin of replication, which is a sequence of DNA capable of directing the propagation of itself and any linked sequence. However, a number of other features are needed and a variety of specialised cloning vectors (small piece of DNA into which a foreign DNA fragment can be inserted) exist that allow protein expression, tagging, single stranded RNA and DNA production and a host of other manipulations.

Cloning of any DNA fragment essentially involves four steps:

1. *Fragmentation*: Breaking apart a strand of DNA
2. *Ligation*: Gluing together pieces of DNA in a desired sequence
3. *Transfection*: Inserting the newly formed pieces of DNA into cells
4. *Screening/selection*: Selecting out the cells that were successfully transfected with the new DNA

Although these steps are invariable among cloning procedures a number of alternative routes can be selected, these are summarized as a 'cloning strategy'.

Initially, the DNA of interest needs to be isolated to provide a DNA segment of suitable size. Subsequently, a ligation procedure is used where the amplified fragment is inserted into a vector (piece of DNA). The vector (which is frequently circular) is linearised using restriction enzymes, and incubated with the fragment of interest under appropriate conditions with an enzyme called DNA ligase. Following ligation the vector with the insert of interest is transfected into cells.

A number of alternative techniques are available, such as chemical sensitivation of cells, electroporation, optical injection and biolistics. Finally, the transfected cells are cultured. As the aforementioned procedures are of particularly low efficiency, there is a need to identify the cells that have been successfully transfected with the vector construct containing the desired insertion sequence in the required orientation. Modern cloning vectors include selectable antibiotic resistance markers, which allow only cells in which the vector has been transfected, to grow.

Additionally, the cloning vectors may contain colour selection markers, which provide blue/white screening (alpha-factor complementation) on X-galmedium. Nevertheless, these selection steps do not absolutely guarantee that the DNA insert is present in the cells obtained. Further investigation of the resulting colonies must be required to confirm that cloning was successful. This may be accomplished by means of PCR, restriction fragment analysis and/or DNA sequencing.

CELL CLONING

Cloning unicellular organisms

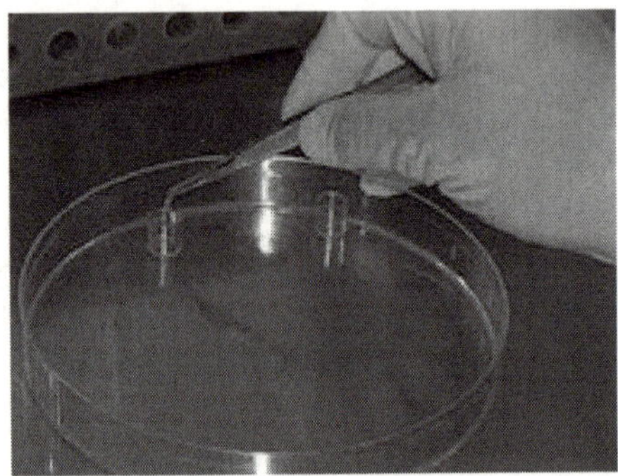

Fig. Cloning Cell-line Colonies Using Cloning Rings

Cloning a cell means to derive a population of cells from a single cell. In the case of unicellular organisms such as bacteria and yeast, this process is remarkably simple and essentially only requires the inoculation of the appropriate medium. However, in the case of cell cultures from multi-cellular organisms, cell cloning is an arduous task as these cells will not readily grow in standard media.

A useful tissue culture technique used to clone distinct lineages of cell lines involves the use of cloning rings (cylinders).According to this technique, a single-cell suspension of cells that have been exposed to a mutagenic agent or drug used to driveselection is plated at high dilution to create isolated colonies; each arising from a single and potentially clonal distinct cell. At an early growth stage when colonies consist of only a few of cells, sterile polystyrene rings (cloning rings), which have been dipped in grease are placed over an individual colony and a small amount of trypsin is added. Cloned cells are collected from inside the ring and transferred to a new vessel for further growth.

Cloning Stem Cells

Somatic-cell nuclear transfer, known as SCNT, can also be used to create embryos for research or therapeutic purposes. The most likely purpose for this is to produce embryos for use in stem cell research. This process is also called "research cloning" or "therapeutic cloning." The goal is not to create cloned human beings (called "reproductive cloning"), but rather to harvest stem cells that can be used to study human development and to potentially treat disease. While a clonal human blastocyst has been created, stem cell lines are yet to be isolated from a clonal source.

Therapeutic cloning is achieved by creating embryonic stem cells in the hopes of treating diseases such as diabetes and Alzheimer's. The process begins by taking out the nucleus (containing the DNA) from an egg cell and putting in it a nucleus from the adult cell to be cloned. In the case of someone with Alzheimer's disease, the nucleus from a skin cell of that patient is placed into an empty egg. The reprogrammed cell begins to develop into an embryo because the egg reacts with the transferred nucleus. The embryo will become genetically identical to the patient. The embryo will then form a blastocyst which has the potential to form/become any cell in the body.

The reason why SCNT is used for cloning is because somatic cells can be easily acquired and cultured in the lab. This process can either add or delete specific genomes of farm animals. A key point to remember is that cloning is achieved when the oocyte maintains its normal functions and instead of using sperm and egg genomes to replicate, the oocyte is inserted into the donor's somatic cell nucleus. The oocyte will react on the somatic cell nucleus, the same way it would on sperm cells.

The process of cloning a particular farm animal using SCNT is relatively the same for all animals. The first step is to collect the somatic cells from the animal that will be cloned. The somatic cells could be used immediately or stored in the laboratory for later use. The hardest part of SCNT is removing maternal DNA from an oocyte at metaphase II. Once this has been done, the somatic nucleus can be inserted into an egg cytoplasm. This creates a one-cell embryo. The grouped somatic cell and egg cytoplasm are then introduced to an electrical current. This energy will hopefully allow the cloned embryo to begin development. The successfully developed embryos are then placed in surrogate recipients, such as a cow or sheep in the case of farm animals.

SCNT is seen as a good method for producing agriculture animals for food consumption. It successfully cloned sheep, cattle, goats, and pigs. Another benefit is SCNT is seen as a solution to clone endangered species that are on the verge of going extinct. However, stresses placed on both the egg cell and the introduced nucleus are enormous, leading to a high loss in resulting cells. For example, the cloned sheep Dolly was born after 277 eggs were used for SCNT, which created 29 viable embryos. Only three of these embryos survived until birth, and only one survived to adulthood. As the procedure currently cannot be automated, and has to be performed manually under amicroscope, SCNT is very resource intensive. The biochemistry involved in reprogramming the differentiated somatic cell nucleus and activating the recipient egg is also far from being well-understood.

In SCNT, not all of the donor cell's genetic information is transferred, as the donor cell's mitochondria that contain their own mitochondrial DNA are left behind. The resulting hybrid cells retain those mitochondrial structures which originally belonged to the egg. As a consequence, clones such as Dolly that are born from SCNT are not perfect copies of the donor of the nucleus.

ORGANISM CLONING

Organism cloning (also called reproductive cloning) refers to the procedure of creating a new multicellular organism, genetically identical to another. In essence this form of cloning is an asexual method of reproduction, where fertilization or inter-gamete contact does not take place. Asexual reproduction is a naturally occurring phenomenon in many species, including most plants and some insects.

Scientists have made some major achievements with cloning, including the asexual reproduction of sheep and cows. There is a lot of ethical debate over whether or not cloning should be used. However, cloning, or asexual propagation, has been common practice in the horticultural world for 100 of years.

Horticultural

The term *clone* is used in horticulture to refer to descendants of a single plant which were produced by vegetative reproduction or apomixis. Many horticultural plant cultivarsare clones, having been derived from a single individual, multiplied by some process other than sexual reproduction. As an example, some European cultivars of grapesrepresent clones that have been propagated for over two millennia.

Other examples are potato and banana. Grafting can be regarded as cloning, since all the shoots and branches coming from the graft are genetically a clone of a single individual, but this particular kind of cloning has not come under ethical scrutiny and is generally treated as an entirely different kind of operation.

Many trees, shrubs, vines, ferns and other herbaceous perennials form clonal colonies naturally. Parts of an individual plant may become detached by fragmentation and grow on to become separate clonal individuals. A common example is in the vegetative reproduction of moss and liverwort gametophyte clones by means of gemmae. Some vascular plants e.g. dandelion and certain viviparous grasses also form seeds asexually, termed apomixis, resulting in clonal populations of genetically identical individuals.

Parthenogenesis

Clonal derivation exists in nature in some animal species and is referred to as parthenogenesis (reproduction of an organism by itself without a mate). This is an asexual form of reproduction that is only found in females of some insects, crustaceans and lizards. The growth and development occurs without fertilization by a male. In plants, parthenogenesis means the development of an embryo from an unfertilized egg cell, and is a component process of apomixis. In species that use the XY sex-determination system, the offspring will always be female. An example is the "Little Fire Ant" (*Wasmannia auropunctata*), which is native to Central and South America but has spread throughout many tropical environments.

Artificial Cloning of Organisms

Artificial cloning of organisms may also be called *reproductive cloning.*

First Moves

Hans Spemann, a German embryologist was awarded a Nobel Prize in Physiology or Medicine in 1935 for his discovery of the effect now known as embryonic induction, exercised by various parts of the embryo, that directs the development of groups of cells into particular tissues and organs. In 1928 he and his student, Hilde Mangold, were the first to perform somatic-cell nuclear transfer using amphibian embryos – one of the first moves towards cloning.

Methods

Reproductive cloning generally uses "somatic cell nuclear transfer" (SCNT) to create animals that are genetically identical. This process entails the transfer of a nucleus from a donor adult cell (somatic cell) to an egg that has no nucleus. If the egg begins to divide normally it is transferred into the uterus of the surrogate mother. Such clones are not strictly identical since the somatic cells may contain mutations in their nuclear DNA. Additionally, the mitochondria in the cytoplasm also contains DNA and during SCNT this mitochondrial DNA is wholly from the cytoplasmic donor's egg, thus the mitochondrial genome is not the same as that of the nucleus donor cell from which it was produced. This may have important implications for cross-species nuclear transfer in which nuclear-mitochondrial incompatibilities may lead to death.

Artificial *embryo splitting* or *embryo twinning,* a technique that creates monozygotic twins from a single embryo, is not considered in the same fashion as other methods of cloning. During that procedure, an donor embryo is split in two distinct embryos, that can then be transferred embryo transfer. It is optimally performed at the 6- to 8-cell stage, where it can be used as an expansion of IVF to increase the number of available embryos. If both embryos are successful, it gives rise to monozygotic (identical) twins.

Obtaining Blastocysts

A blastocyst is formed in the early stage of the development of an embryo. During cloning process, the blastocyst cells often are obtained by scientists 5 days after the egg has been divided.

Dolly the Sheep

Dolly, a Finn-Dorset ewe, was the first mammal to have been successfully cloned from an adult cell. Dolly was formed by taking a cell from the udder of her biological mother. Her biological mother was 6 years old when the cells were taken from her udder.Dolly's embryo was created by taking the cell and inserting it into a sheep ovum.

It took 434 attempts before an embryo was successful. The embryo was then placed inside a female sheep that went through a normal pregnancy. She was cloned at the Roslin Institute in Scotland and lived there from her birth in 1996 until her death in 2003 when she was six. She was born on July 5, 1996 but not announced to the world until February 22, 1997. Her stuffed remains were placed at Edinburgh's Royal Museum, part of the National Museums of Scotland.

Dolly was publicly significant because the effort showed that genetic material from a specific adult cell, programmed to express only a distinct subset of its genes, can be reprogrammed to grow an entirely new organism. Before this demonstration, it had been shown by John Gurdon that nuclei from differentiated cells could give rise to an entire organism after transplantation into an enucleated egg. However, this concept was not yet demonstrated in a mammalian system.

Cloning Dolly the sheep had a low success rate per fertilized egg; she was born after 277 eggs were used to create 29 embryos, which only produced three lambs at birth, only one of which lived. Seventy calves have been created and one third of them died young; Prometea took 814 attempts. Notably, although the first clones were frogs, no adult cloned frog has yet been produced from a somatic adult nucleus donor cell.

There were early claims that Dolly the Sheep had pathologies resembling accelerated aging. Scientists speculated that Dolly's death in 2003 was related to the shortening of telomeres, DNA-protein complexes that protect the end of linear chromosomes. However, other researchers, including Ian Wilmut who led the team that successfully cloned Dolly, argue that Dolly's early death due to respiratory infection was unrelated to deficiencies with the cloning process. Dolly was named after Dolly Parton because the cells cloned to make her were from an udder.

Species cloned

The modern cloning techniques involving nuclear transfer have been successfully performed on several species.

Notable experiments include:

- *Tadpole*: (1952) Robert Briggs and Thomas J. King had successfully cloned northern leopard frogs: thirty-five complete embryos and twenty-seven tadpoles from one-hundred and four successful nuclear transfers.
- *Carp*: (1963) In China, embryologist Tong Dizhou produced the world's first cloned fish by inserting the DNA from a cell of a male carp into an egg from a female carp. He published the findings in a Chinese science journal.
- *Mice*: (1986) A mouse was successfully cloned from an early embryonic cell. Soviet scientists Chaylakhyan, Veprencev, Sviridova, and Nikitin had the mouse "Masha" cloned. Research was published in the magazine "Biofizika" volume ÕÕÕII, issue 5 of 1987.
- *Sheep*: Marked the first mammal being cloned (1984) from early embryonic cells by Steen Willadsen. Megan and Morag cloned from differentiated embryonic cells in June 1995 and Dolly the sheep from a somatic cell in 1996.
- *Rhesus Monkey*: Tetra (January 2000) from embryo splitting
- *Gaur*: (2001) was the first endangered species cloned.
- *Cattle*: Alpha and Beta (males, 2001) and (2005) Brazil
- *Cat*: CopyCat "CC" (female, late 2001), Little Nicky, 2004, was the first cat cloned for commercial reasons
- *Rat*: Ralph, the first cloned rat (2003)
- *Mule*: Idaho Gem, a john mule born 4 May 2003, was the first horse-family clone.
- *Horse*: Prometea, a Haflinger female born 28 May 2003, was the first horse clone.
- *Dog*: Snuppy, a male Afghan hound was the first cloned dog (2005).
- *Wolf*: Snuwolf and Snuwolffy, the first two cloned female wolves (2005).
- *Water Buffalo*: Samrupa was the first cloned water buffalo. It was born on February 6, 2009, at India's Karnal National Diary Research Institute but died five days later due to lung infection.
- Pyrenean Ibex (2009) was the first "extinct" animal (while the species is not extinct, nor even endangered, no living examples of the Pyrenean subspecies had been known since 2000) to be cloned back to life; the clone lived for seven minutes before dying of lung defects.
- *Camel*: (2009) Injaz, is the first cloned camel.
- *Pashmina goat*: (2012) Noori, is the first cloned pashmina goat. Scientists at the faculty of veterinary sciences and animal husbandry of Sher-e-Kashmir University of Agricultural Sciences and Technology of Kashmir successfully cloned the first Pashmina goat (Noori) using the advanced reproductive techniques under the leadership of Riaz Ahmad Shah.

Human Cloning

Human cloning is the creation of a genetically identical copy of an existing or previously existing human. The term is generally used to refer to *artificial* human cloning; human clones in the form of identical twins are commonplace, with their cloning occurring during the natural process of reproduction. There are two commonly discussed types of human cloning: therapeutic cloning and reproductive cloning.

Therapeutic cloning involves cloning adult cells for use in medicine and is an active area of research. Reproductive cloning would involve making cloned humans. A third type of cloning called replacement cloning is a theoretical possibility, and would be a combination of therapeutic and reproductive cloning. Replacement cloning would entail the replacement of an extensively damaged, failed, or failing body through cloning followed by whole or partial brain transplant.

The various forms of human cloning are controversial. There have been numerous demands for all progress in the human cloning field to be halted. Most scientific, governmental and religious organizations oppose reproductive cloning. The American Association for the Advancement of Science (AAAS) and other scientific organizations have made public statements suggesting that human reproductive cloning be banned until safety issues are resolved. Serious ethical concerns have been raised by the future possibility of harvesting organs from clones.

Some people have considered the idea of growing organs separately from a human organism - in doing this, a new organ supply could be established without the moral implications of harvesting them from humans. Research is also being done on the idea of growing organs that are biologically acceptable to the human body inside of other organisms, such as pigs or cows, then transplanting them to humans, a form of xenotransplantation.

The first hybrid human clone was created in November 1998, by Advanced Cell Technologies. It was created from a man's leg cell, and a cow's egg whose DNA was removed. It was destroyed after 12 days. Since a normal embryo implants at 14 days, Dr Robert Lanza, ACT's director of tissue engineering, told the Daily Mail newspaper that the embryo could not be seen as a person before 14 days. While making an embryo, which might have resulted in a complete human had it been allowed to come to term, according to ACT: "[ACT's] aim was 'therapeutic cloning' not 'reproductive cloning'"

On January, 2008, Wood and Andrew French, Stemagen's chief scientific officer in California, announced that they successfully created the first 5 mature human embryos using DNA from adult skin cells, aiming to provide a source of viable embryonic stem cells. Dr. Samuel Wood and a colleague donated skin cells, and DNA from those cells was transferred to human eggs. It is not clear if the embryos produced would have been capable of further development, but Dr. Wood stated that if that were possible, using the technology for reproductive cloning would be both unethical and illegal. The 5 cloned embryos, created in Stemagen Corporation lab, in La Jolla, were destroyed.

Ethical Issues of Cloning

Because of recent technological advancements, the cloning of animals (and potentially humans) has been an issue. Many religious organizations oppose all forms of cloning. Judaism does not equate life with conception and, though some question the wisdom of cloning, Orthodox Judaism rabbis generally find no firm reason inJewish law and ethics to object to cloning. From the standpoint of classical liberalism, concerns also exist regarding the protection of the identity of the individual and the right to protect one's genetic identity.

Gregory Stock is a scientist and outspoken critic against restrictions on cloning research. The social implications of an artificial human production scheme were famously explored in Aldous Huxley's novel *Brave New World.* On December 28, 2006, the U.S. Food and Drug Administration (FDA) approved the consumption of meat and other products from cloned animals. Cloned-animal products were said to be virtually indistinguishable from the non-cloned animals. Furthermore, companies would not be required to provide labels informing the consumer that the meat comes from a cloned animal.

Critics have raised objections to the FDA's approval of cloned-animal products for human consumption, arguing that the FDA's research was inadequate, inappropriately limited, and of questionable scientific validity. Several consumer-advocate groups are working to encourage a tracking program that would allow consumers to become more aware of cloned-animal products within their food.

Joseph Mendelson, legal director of the Center for Food Safety, said that cloned food still should be labeled since safety and ethical issues about it remain questionable. Carol Tucker Foreman, director of food policy at the Consumer Federation of America, stated that FDA does not consider the fact that the results of some studies revealed that cloned animals have increased rates of mortality and deformity at birth. Another concern is that the biotechnologies used on animals may someday be used on humans. Some people may be more open to the idea of cloning of animals because most western countries have passed legislation against cloning humans, yet only a few countries passed legislation against cloning animals.

Possible Abnormalities Due to Cloning

Researchers have found several abnormalities in cloned organisms, particularly in mice. The cloned organism may be born normal and resemble its non-cloned counterpart, but majority of the time will express changes in its genome later on in life. The concern with cloning humans is that the changes in genomes may not only result in changes in appearance, but in psychological and personality changes as well. The theory behind this is that the biological blueprint of the genes is the same in cloned animals as it is in normal ones, but they are read and expressed differently. DNA arrays were used to prove this claim in the research lab of Professor Rudolf Jaenisch. Jaenisch studied placentas from cloned mice and found that one in every 25 genes was expressed abnormally. Results of these abnormally expressed genes in the cloned mice were premature death, pneumonia, liver failure and obesity.

Cloning Extinct and Endangered Species

Cloning, or more precisely, the reconstruction of functional DNA from extinct species has, for decades, been a dream of some scientists. The possible implications of this were dramatized in the best-selling novel by Michael Crichton and high budget Hollywood thriller *Jurassic Park*. In real life, one of the most anticipated targets for cloning was once the Woolly Mammoth, but attempts to extract DNA from frozen mammoths have been unsuccessful, though a joint Russo-Japanese team is currently working toward this goal. And in January 2011, it was reported by Yomiuri Shimbun that a team of scientists headed by Akira Iritani of Kyoto University had built upon research by Dr. Wakayama, saying that they will extract DNA from a mammoth carcass that had been preserved in a Russian laboratory and insert it into the egg cells of an African elephant in hopes of producing a mammoth embryo. The researchers said they hoped to produce a baby mammoth within six years.

In 2001, a cow named Bessie gave birth to a cloned Asian gaur, an endangered species, but the calf died after two days. In 2003, a banteng was successfully cloned, followed by three African wildcats from a thawed frozen embryo. These successes provided hope that similar techniques (using surrogate mothers of another species) might be used to clone extinct species. Anticipating this possibility, tissue samples from the last *bucardo* (Pyrenean Ibex) were frozen in liquid nitrogen immediately after it died in 2000. Researchers are also considering cloning endangered species such as the giant panda and cheetah. The "Frozen Zoo" at the San Diego Zoo now stores frozen tissue from the world's rarest and most endangered species.

In 2002, geneticists at the Australian Museum announced that they had replicated DNA of the Thylacine (Tasmanian Tiger), at the time extinct for about 65 years, usingpolymerase chain reaction. However, on February 15, 2005 the museum announced that it was stopping the project after tests showed the specimens' DNA had been too badly degraded by the (ethanol) preservative. On 15 May 2005 it was announced that the Thylacine project would be revived, with new participation from researchers in New South Wales and Victoria.

In January 2009, for the first time, an extinct animal, the Pyrenean ibex mentioned above was cloned, at the Centre of Food Technology and Research of Aragon, using the preserved DNA of the skin samples from 2001 and domestic goat egg-cells. (The ibex died shortly after birth due to physical defects in its lungs.) One of the continuing obstacles in the attempt to clone extinct species is the need for nearly perfect DNA. Cloning from a single specimen could not create a viable breeding population in sexually reproducing animals. Furthermore, even if males and females were to be cloned, the question would remain open whether they would be viable at all in the absence of parents that could teach or show them their natural behavior.

Cloning endangered species is a highly ideological issue. Many conservation biologists and environmentalists vehemently oppose cloning endangered species—mainly because they think it may deter donations to help preserve natural habitat and wild animal populations. The "rule-of-thumb" in animal conservation is that, if it is still feasible to conserve habitat and viable wild populations, breeding in captivity should not be undertaken in isolation. In a 2006 review, David Ehrenfeld concluded that cloning in animal conservation is an experimental technology that, at its state in 2006, could not be expected to work except by pure chance and utterly failed a cost-benefit analysis. Furthermore, he said, it is likely to siphon funds from established and working projects and does not address any of the issues underlying animal extinction (such as habitat destruction, hunting or other overexploitation, and an impoverished gene pool). While cloning technologies are well-established and used on a regular basis in plant conservation, care must be taken to ensure genetic diversity. He concluded:

Vertebrate cloning poses little risk to the environment, but it can consume scarce conservation resources, and its chances of success in preserving species seem poor. To date, the conservation benefits of transgenics and vertebrate cloning remain entirely theoretical, but many of the risks are known and documented. Conservation biologists should devote their research and energies to the established methods of conservation, none of which require transgenics or vertebrate cloning.

On the 7 December 2011 it was announced that a team from the Siberian mammoth museum and Japan's Kinki University plan to clone a woolly mammoth from a well preserved sample of bone marrow found in August 2011. The team claim that the cloning could be complete within the next five years. Others have expressed doubt about the feasibility of the experiment. Scientists at the University of New South Wales announced in March 2013 that the very recently extinct gastric-brooding frog would be the subject of a cloning attempt to resurrect the species.

IN SCIENCE FICTION

Science fiction has used cloning, most commonly and specifically human cloning, due to the fact that it brings up controversial questions of identity. In Aldous Huxley's*Brave New World* (1932), human cloning is a major plot device that not only drives the story but also makes the reader think critically about what identity means; this concept was re-examined fifty years later in C. J. Cherryh's novels *Forty Thousand in Gehenna* (1983) and *Cyteen* (1988). Kazuo Ishiguro's 2005 novel *Never Let Me Go* centres on human clones and considers the ethics of the practice. Another book that embodies the ideas of cloning is *The House of the Scorpion* which explores the rights of human clones and organ harvesting, set from the eyes of a clone. The short novel *Containing God* by Dr.S.M.Wasi Haider similarly deals with the ideas of cloning and the ethics, lust, and other issues revolving around the topic, emphasizing the idea that creating life gives people the false sense of divinity.

Star Wars portrays human cloning in *Star Wars Episode II: Attack of the Clones* and *Star Wars Episode III: Revenge of the Sith*, in the form of the Grand Army of the Republic, an army of cloned soldiers. The Expanded Universe also has numerous examples of cloning, including the Thrawn trilogy, The Hand of Thrawn duology, and Clone Wars-era media. Popular sci-fi movies that also feature cloning include *Jurassic Park (film), The Island (2005 film),* and *Resident Evil (film series).*

Cloning is also featured in the *Halo* franchise, particularly a technique known as "flash cloning" in which the unstable clone of an individual is created in an incredibly short span of time. Flash cloning is used by the UNSC to kidnap young children for induction into the SPARTAN-II military program, who are surreptitiously replaced by flash clones which die within a short span of time to ensure that no one looks for the children. In the anime *A Certain Scientific Railgun,* Level 5 esper Mikoto Misaka was cloned commercially over 20000 times for the purposes of research into the possibility of a "Level 6" esper.

The book *Cloud Atlas* has science-fiction chapter in the style of an interview with a clone called Sonmi~451, who was born in a futuristic, dystopian Seoul. She is one of thousands of clones created for manual and emotional labor; Sonmi herself works as a server in a restaurant. She later discovers that the sole source of food for clones, called 'Soap', is manufactured from the clones themselves. In 2012, a Japanese television show named "Bunshin" was created. The story's main character, Mariko, is a woman studying child welfare in Hokkaido. She grew up always doubtful about the love from her mother, who looked nothing like her and who died nine years before. One day, she finds some of her mother's belongings at a relative's house, and heads to Tokyo to seek out the truth behind her birth. She later discovered that she was a clone In the 2013 television show *Orphan Black,* cloning is used as a scientific double blind study on the behavioral adaptation of the clones.

THERAPEUTIC CLONING

Nuclear cloning, which has also been called nuclear transplantation and nuclear transfer, involves the introduction of a nucleus from a donor cell into an enucleated oocyte to generate an embryo with a genetic makeup identical to that of the donor. Although there has been tremendous interest in the field of nuclear cloning since the birth of Dolly in 1997, the first successful nuclear transfer was reported over 50 yr ago by Briggs and King. Cloned frogs, which were the first vertebrates derived from nuclear transfer, were subsequently reported by Gurdon in 1962, but the nuclei were derived from nonadult sources.

In the past 6 yr, tremendous advances in nuclear cloning technology have been reported, indicating the relative immaturity of the field. Dolly was not the first cloned mammal to be produced from adult cells; in fact, live lambs were produced in 1996 using nuclear transfer and differentiated epithelial cells derived from embryonic discs. The significance of Dolly was that she was the first mammal to be derived from an adult somatic cell using nuclear transfer. Since then, animals from several species have been grown using nuclear transfer technology, including cattle, goats, mice, and pigs.

Two types of nuclear cloning, reproductive cloning and therapeutic cloning, have been described, and a better understanding of the differences between the two types may help to alleviate some of the controversy that surrounds these revolutionary technologies. Banned in most countries for human applications, reproductive cloning is used to generate an embryo that has the identical genetic material as its cell source. This embryo can then be implanted into the uterus of a female to give rise to an infant that is a clone of the donor.

On the other hand, therapeutic cloning is used to generate early stage embryos that are explanted in culture to produce embryonic stem cell lines whose genetic material is identical to that of its source. These autologous stem cells have the potential to become almost any type of cell in the adult body, and thus would be useful in tissue and organ replacement applications. Some useful applications would be in the treatment of diseases, such as end-stage kidney disease, neurodegenerative diseases, and diabetes, for which there is limited availability of immunocompatible tissue transplants.

Therefore, therapeutic cloning, which has also been called somatic cell nuclear transfer, provides an alternative source of transplantable cells that theoretically may be limitless. The strategy of combining therapeutic cloning with tissue engineering to develop tissues and organs. According to data from the Centers for Disease Control and Prevention, it has been estimated that approximately 3000 Americans die every day of diseases that could have been treated with embryonic stem cell–derived tissues.

With current allogeneic tissue transplantation protocols, rejection is a frequent complication because of immunologic incompatibility, and immunosuppressive drugs are usually administered to treat and hopefully prevent graft-*versus*-host disease. The use of transplantable tissue and organs derived from therapeutic cloning may lead to the avoidance of immune responses that typically are associated with transplantation of nonautologous tissues. As a result, with therapeutic cloning, the variety of serious and potentially life-threatening complications associated with immunosuppressive treatments may be avoided.

THERAPEUTIC CLONING: PROMISES AND ISSUES

The advancement in biotechnologies and stem cell research, although encountering many scientific difficulties, legal constraints and ethical roadblocks, offers a tremendous potential in regenerative medicine and in the treatment of genetic defects. Therapeutic cloning is the transfer of nuclear material isolated from a somatic cell into an enucleated oocyte in the goal of deriving embryonic cell lines with the same genome as the nuclear donor. Somatic cell nuclear transfer (SCNT) products have histological compatibility with the nuclear donor, which circumvents, in clinical applications, the use of immunosuppressive drugs with heavy side-effects.

While the goal of reproductive cloning is the creation of a person, the purpose of therapeutic cloning is to generate and direct the differentiation of patient-specific cell lines isolated from an embryo not intended for transfer in utero. Therapeutic cloning, through the production of these autologous nuclear-transfer embryonic stem cells (ntESC), offers great promises for regenerative and reproductive medicine, and in gene therapy, as a vector for gene-delivery. This review focuses on the recent breakthroughs in research based on therapeutic cloning, their feasibility, and their potential applications in medicine. The second part of this review discusses current roadblocks of therapeutic cloning, both in science and biomedical ethics, as well as the main alternatives to therapeutic cloning.

Procedure for SCNT and Characteristics of the ntESC

The procedure for SCNT does not differ from that of reproductive cloning. The host oocyte is arrested at metaphase II, and immobilized through light suction exerted by a pipette tip. A glass needle is used to remove a small piece of the zona pellucida and is reinserted through this puncture to extract the polar body and the oocyte nuclei. The incorporation of the somatic nuclei into the enucleated oocyte can be done through electrofusion, which is the application of an electric pulse to incorporate a mammalian cell into the oocyte (used to produce Dolly). Alternatively, a somatic nucleus can be injected in the perivitelline space, the fluid-filled region between the zona pellucida and the ooplasm, as was used for Cumulina, the first mouse cloned through SCNT.

Mitosis occurs in vitro until the formation of the blastocyst, a fluid-filled hollow ball of cells (40–150 cells) to which is attached, from the inside, the embryoblast or inner cell mass from which ntESC are taken. Subsequent addition of cell-type specific markers and growth hormones promotes the differentiation of the ntESC into the desired cell-line to be implanted in vivo inside the nuclear donor for therapeutic purposes, in cell replacement therapy for instance. In vitro, the ESC can proliferate ad infinitum and are totipotent, capable of differentiating into any cell-type of the body, contrary to adult stem cells which are multipotent, namely committed to produce any type of cells pertaining to a particular lineage.

Current Legal Status of Therapeutic Cloning in Relation to Reproductive Cloning

Laws regarding biomedicine are generally formulated in vague terms that do not distinguish reproductive from therapeutic cloning. The Convention on Human Rights and Biomedicine (Oviedo convention), formulated by the Council of Europe in 1997, is counterintuitive. Article 13 declares that "an intervention seeking to modify the human genome may be only undertaken for preventive, diagnostic or therapeutic purposes" and stipulates in Article 18 that "the creation of human embryos for research is prohibited." The Protocol on Cloning, put forward in 1998 and signed by 19 European nations, bans reproductive cloning and was paradoxically signed by France and Germany, which both have permissive policies regarding the generation of human ntESC lines.

Committees are formed in different countries to debate and regulate cloning, such as the President's Council on Bioethics, created in the USA in 2002, which is a much less permissive group than the UK's Human Fertilization and Embryology Authority (HFEA). The legitimacy of the latter is being questioned by the Prolife movement under the pretext that they were not democratically elected.

Canada's Assisted Human Reproduction Act, in vigor since 2004, allows stem cell research only on unimplanted embryos obtained from fertility clinics but forbids SCNT. Asia has the highest legal permissibility since the generation of human ntESC lines through SCNT is legal. Australia is currently reviewing its existing laws to follow the Asian trend in Singapore, China and South Korea, and to legalize the generation of chimeras using human DNA.

Since both reproductive and therapeutic cloning require the in vitro generation of a human embryo, prohibiting reproductive cloning is likely to result in severely hindering medically important research based on therapeutic cloning. A worldwide ban on reproductive human cloning was proposed by France and Germany to the UN in 2001 and effective4 since September 2006. A breakthrough in reproductive cloning was published a month earlier by Zavos and Illmensee, who injected a skin fibroblast nucleus from an infertile man into an oocyte provided by his wife.

One out of three SCNT attempts was successful, and although the four-celled embryo failed to implant in utero, this is the "first evidence of the creation and transfer of a human cloned embryo for reproductive purposes." One may infer, from the rigidity of the current legislature regarding therapeutic cloning and stem cell research, that legal constraints are motivated by the fear that scientific development will be faster than the legislative debate, which was almost the case with Zavros and Illmensee's breakthrough, and lead to the unregulated reproductive cloning of human beings.

Promises of Therapeutic Cloning

SCNT in the context of therapeutic cloning holds a huge potential for research and clinical applications including the use of SCNT product as a vector for gene delivery, the creation of animal models of human diseases, and cell replacement therapy in regenerative medicine. Furthermore, SCNT might, in the future, allow in vitro organogenesis and counteract senescence. The combination of therapeutic cloning and gene therapy offers a great potential for patient-specific rescue of a genetic mutation of the loss-of-function type, resulting in lowered or eliminated activity of a particular protein.

Therapeutic cloning used in cell replacement therapy has the potential to create various types of tissues such as osteoblasts to counteract osteoporosis, and spinal cord regeneration following trauma, as shown by Deshpande et al, who transferred motor neurons derived from ESC to rats with a severed spinal cord. The resulting recovery of motility could lead to clinical applications for paralysis in humans through therapeutic cloning.

Applications in Regenerative Medicine: Recent Breakthroughs for Diabetes and Neurodegenerative Diseases

Therapeutic cloning constitutes a promising tool in tissue engineering and might offer the possibility of synthesizing organs de novo, which would solve the problems of immune rejection and organ shortage for transplantation. The assembly of patient-specific cardiomyocytes, blood vessels and skin pieces fixed on a scaffold holds great hope in the treatment of infarctus, atherosclerosis and severe burns, respectively.

Consequently, the feasibility of de novo organogenesis based on SCNT depends on the elucidation of the tissue-specific molecular pathways mediating differentiation as well as the improvement of current SCNT and tissue engineering methods in order to recreate in vitro the complex three-dimensional organization and different intercellular interactions in organogenesis.

D'Amour et al designed in 2006 a five-step protocol enabling human embryonic stem cells to differentiate into endocrine cells producing most pancreatic hormones, including glucagon and insulin, with implications for use in cell replacement therapy for the treatment of diabetes mellitus. The combination of growth factors and differentiation markers added at each stage of the method were designed to duplicate pancreatic organogenesis in vivo, a breakthrough which could concretize the hope of generating organs using therapeutic cloning.

According to this protocol, patient-specific ntESC lines would be differentiated into successive cell-type intermediates representative of pancreatic organogenesis, from endoderm to the terminally differentiated â-cell that would then be transplanted into the patient's pancreas to treat diabetes-related hyperglycemia.

The recent success of therapeutic cloning in a mouse model of Parkinson's disease foreshadows clinical applicability in humans to treat neurodegenerative diseases and conditions involving demyelination. Parkinson's disease is characterized by the deterioration of dopaminergic neurons resulting in constant tremor and muscular stiffness impairing motility. Barberi et al derived, by SCNT with somatic nuclei from mouse cumulus and tail-tip cells, two ntESC lines which were induced to differentiate into motor, GABAminergic, serotonergic and dopaminergic neurons forming synapses and displaying normal electrophysiological properties in vitro.

The dopaminergic neurons were directly injected into the cortical striatum of mice with Parkinson-like lesions induced by 6-hydroxydopamine. Long-term behavioral rescue was observed, and 80 per cent of the ntESC derived neurons were alive 8 week post-transplantation, contrary to only 40 per cent for stem cell-derived neurons. Hence, the therapeutic cloning approach was shown to be more permanent as a cell replacement therapy and could eventually be extended to the treatment of cortical atrophy resulting from stroke or Alzheimer's disease.

REJUVENATING POTENTIAL

It has been observed that the SCNT of the nucleus of a cell close to reaching senescence resets the lifespan of the cell, as seen by the ability of the resulting embryo to carry 31 rounds of division, compared to 33 for a wild-type embryo of the same developmental stage. The rejuvenating potential conferred by SCNT can be paradoxically thwarted by telomere shortening, which reflects both the biological age of the nuclear donor and the time during which the ntESC lines were grown, leading to premature aging also observed in cloned animals.

However, the addition of a transgene containing the two coding regions needed for the production of telomerase could restore telomere lengths and thus increase the survival of the transplanted cells, which would increase the success rate of therapeutic cloning for regenerative medicine. This strategy needs more investigation to be feasible because, while telomere length would not trigger tumorigenesis, Stampfer et al showed that, knockout p16INK4a epithelial cells with no endogenous telomerase activity do not respond to the proapoptotic signals of the growth factor TGF-â following the addition of high levels of hTERT, the telomerase catalytic subunit.

GENERATING ANIMAL MODELS OF HUMAN DISEASE

Animal models of human diseases can be designed through therapeutic cloning for research purposes. Although a viable nonhuman primate has not yet been produced by SCNT, the success of Mitalipov and Wolf in creating a monkey by embryonic cloning from the nucleus of an allogenic blastomere supported the possibility that, through gene targeting, genetic defects can be reproduced in a wild-type genome to express a loss of function. Hence, we are getting one step closer to patient-specific genetic engineering of animal models of human disease. With improved SCNT protocol, the nucleus of a patient's skin biopsy could be introduced into a primate or mouse enucleated oocyte so that the resulting clone expresses the condition in a patient-specific way. Hence, clinical testing would be done on the animal model to find an optimal treatment, such as the drug combination to treat epigenetically-triggered cancer, highly variable among instances.

CANCER DIAGNOSIS

SCNT has applications in cancer research to identify whether a particular type of cancer arises from a genetic or an epigenetic defect, such as the demethylation of a tumor suppressor gene. The epigenetic modifications of chromatin structure in cancerous cells involve altered histone methylation, phosphorylation and deacetylation, as well as DNA methylation, which are reversible unlike genetic mutations. Supportive evidence for oncogenesis resulting from epigenetic features includes studies where normal mice blastula were obtained through SCNT from a skin malignancy and a medullar tumor.

These studies could lead to clinical applications for cancer diagnosis in humans since nuclear reprogramming signals from the host ooplasm variably reset the epigenetic profile of the nuclear donor DNA. The derivation through SCNT of a healthy patient-specific stem line would show that cancer onset was triggered by epigenetic alterations. Anti-methylation drug, such as 5-aza-22 -deoxycytidine inhibiting DNA methyltransferases that inactivate apoptotic genes in cancerous cells and histone deactylase inhibitors against oncogene overexpression are currently under clinical trial as a potential anti-cancer therapy.

However, epigenetic resetting following SCNT is likely to disrupt normal phenotype of the embryo-derived cell lines and the adult clone, the latter displaying an abnormally low body weight and expression level of MUP encoding genes (Major Urinary Proteins) as shown by Reik et al in the mouse.

The epigenetic pattern of imprinted genes that was established during gametogenesis is lost through SCNT and the inactivation of early genes directing embryogenesis can explain low embryo viability and poor efficiency in the derivation of autologous ntESC lines.

Blelloch et al found out, from studies on neurons, that stem cells used as the nuclear donor have a higher success rate than fully differentiated cells in the derivation of autologous embryonic cells. The introduction of a genetic mutation to reduce the function of DNA methyltransferase-1, as investigated by the same team, improved the production of ntESC due to resulting "global hypomethylation" of genomic nuclear DNA.

THERAPEUTIC CLONING IN THE CONTEXT OF GENE THERAPY: CURRENT HOPES AND DRAWBACKS

SCNT from genetically modified nuclei obtained from a patient's skin biopsy, for example, is an efficient strategy to restore normal expression of a missing factor or to facilitate in vivo survival of the graft generated. For instance, patient-specific cardiomyocytes produced through SCNT will not integrate into the scarred heart tissue resulting from myocardial infarction. A proposed strategy would be the genomic integration of an exogenous gene, prior to transplantation, encoding an anti-scarring factor such as TGF-β transforming growth factor-β). In the case of haemophilia, characterized by a deficit in functional clotting factor IX and XIII, the addition of the genetically engineered missing DNA sequence, or the replacement of the dysfunctional gene through homologous recombination in a patient's biopsy prior to SCNT could produce patient-specific cell lines with a correction for the defect.

For instance, Duchenne Muscular Dystrophy (DMD) is an inheritable X-linked condition characterized by reduced intramuscular dystophin levels, causing cellular necrosis and weakening. Being a single-gene disorder, DMD can be treated by therapeutic cloning in combination with gene therapy to restore normal dystrophin production. In the case where ntESC are transplanted without prior differentiation in vitro, the insertion of a transgene encoding MyoD, a transcription factor responsible for commitment to the myogenic lineage, may promote muscle regeneration.

The combination of gene therapy and therapeutic cloning has exciting potential for the genetic rescue of missing alleles in heritable genetic disorders such as severe combined immunodeficiency (SCID), in which genetic mutations of specific genes such as RAG-1 and 2, essential for the DNA recombination allowing immunoglobulin and lymphocyte polymorphism, render the immune system completely inefficient. Hochedlinger et al took a somatic nucleus from the tail-tip of an SCID mouse-model, created through the double-knockout of the Rag-2 gene (recombination-activating gene 2), and rescued the genetic defect through the insertion of two copies of the Rag-2 gene by homologous recombination. SCNT was performed to clone viable Rag-2 (+/+) mice with a normal immune system, from which embryonic stem cells were differentiated in vitro into hematopoietic stem cells normally found in the bone marrow.

Three weeks following transplantation into the Rag-2(–/–) knockout mouse model, partial rescue of immune function was observed, as well as the presence of lymphocyte precursors and functional antibodies. However, mature T lymphocytes were not observed, suspected to be due to selective differentiation of the transplanted stem cells into myeloid cells (bone marrow precursors) instead. Although more work needs to be done to elucidate the pathways leading to preferential differentiation in vivo, the combination of gene therapy for the rescue of a loss of function and therapeutic cloning to bypass graft rejection holds the potential to eventually cure other immune disorders. Oncogenic activation following transduction constitutes a major drawback to this approach. In 2002, the insertion of the transgene to treat X-linked SCID in the LMO2 oncogene caused the onset of leukemia in two out of seven patients recently treated. Repeated graft rejection, even when derived through SCNT, remains an unsolved problem in the case of autoimmune disorders such as pernicious anemia and multiple sclerosis.

ROADBLOCKS OF THERAPEUTIC CLONING

Legal and Funding Issues

Legislative constraints and the subsequent lack of funding constitute a major impediment to the advancement of therapeutic cloning. For instance, although therapeutic cloning is not completely banned in the United States, federal funding is not permitted to be used in experiments involving the 20 cell lines in the NIH (National Institute of Health) registry derived before August 9, 2001. Out of these cell lines approved by Bush, 12 died and the remaining is not useful for research purposes.

Researchers have to therefore rely on the scarcity of private funding, although 4 American states, including California with a yearly investment of 295 millions, have a budget allowed specifically for stem cell research. Clinton expressed concern on chimera production through "experiments involving the mingling of human and non-human species" but paradoxically all the cell lines in the NIH registry that got Bush's approval were grown on animal-feeder cell layers and therefore contain traces of animal contaminants.

Oocyte Availability: Regulations and Ethical Concerns

A major roadblock in the feasibility of human therapeutic cloning is the low availability of oocytes for research purposes. Currently, due to low SCNT efficiency, it is estimated that 280 human oocytes would be needed in order to derive one observe patient-specific ntESC line. The Human Fertility and Embryo Authority, in England, allows women in fertility clinics to offer two oocytes for scientific research, provided that at least twelve oocytes are collected, although the extra oocytes taken are more likely to be donated for in vitro insemination. Oocyte donations are, by law, forbidden to be remunerated other than to reimburse the cost of the procedure (from 1000 to 2000$) for ethical reasons.

Substantial financial gain would incite poorer women to surrender of part of a finite supply of gametes, in addition to the risks incurred through surgerical removal of the oocytes and hormonal treatments. Ovarian hyperstimulation syndrome (OHSS) results, in most cases, from the administration of drugs such as gonadotropin-releasing hormone agonists, to induce the simultaneous maturation of multiple follicles into oocytes. OHSS can lead to cardio-respiratory difficulties, renal problems and internal hemorrhage, due to the accumulation of fluid in the abdominal cavity caused by abnormal vascular permeability in the vicinity of the ovaries, and occurs in 2 to 5 per cent of the patients.

Fertility clinics are the major source of human oocytes for research. The aged oocytes that did not fertilize during in vitro trials are not optimal for SCNT, as investigated by Hall et al, who observed overexpression of genes encoding for meiotic spindles proteins and a lower cleavage efficiency of aged oocytes versus fresh ones. After two years of debate, Harvard is the only group currently allowed to use oocytes collected for the sole purpose of research, with informed consent. Eggan and Melton intend to generate human ntESC lines from patients with diabetes, sickle cell anemia and amyotrophic lateral sclerosis (ALS), a progressive neurodegenerative condition targeting motor neurons.

Possible Solutions to the Oocyte Shortage for Therapeutic Cloning

Interestingly, SCNT could provide a solution to low human oocyte availability and a promising therapeutic approach to circumvent infertility. As reported by Nagy and Chang, artificial gametes can be created by haploidization, through SCNT into an enucleated oocyte ready to undergo meiosis upon induction. Tesarik et al incorporated the nucleus of a human cumulus cell into an enucleated allogenic oocyte. They got a 50 per cent success rate in haploidization, and two out of the six artificial oocytes were successfully fertilized in vitro and thawed for possible implantation later on. However, abnormal chromosomal segregation and mitotic spindle assembly, as observed in mice, need to be resolved before haploidization through SCNT can be safely wide-spread in fertility clinics.

Once the molecular pathways of oocyte maturation are resolved, immature follicles could be collected postmortem and induced to mature in vitro for research purpose. However, this option needs to be rigorously regulated, and is likely to stir an ethical controversy.

An alternative to low human oocyte availability would be to use an oocyte of a different species. Successful trials were done to generate blastocysts in vitro through the SCNT of skin fibroblast nucleus from different mammals (ungulates, rodents, pigs, monkey) and human into an enucleated bovine oocyte.

However, only 1 per cent of the human-bovine hybrids developed beyond the 16-cell stage, allowing the derivation of a human ntESC line. The insertion of human nuclear genome into an animal oocyte, especially through electrofusion where both human and animal mitochondrial DNA coexist in the same ooplasm, raise objections among the detractors of therapeutic cloning, although the percentage of total residual animal DNA (nuclear and mitochondrial) is too low to consider the hybrid as a chimera.

Mitochondrial Heteroplasmy

Immune rejection of the ntESC in cell replacement therapy is due to mitochondrial heteroplasmy as a consequence of SCNT since the nuclear donor and ooplasmic host cells are not autologous in most cases. Mitochondrial heteroplasmy is also a major cause of SCNT embryo inviability beyond the eight-cell stage because the mitochondrial-nucleus interactions necessary for the production of most mitochondrial proteins are disrupted due to inter-species incompatibility. Also, antigens such as Mta are encoded by the mitochondrial genome and trigger an autoimmune response targeting the hybrid after transplantation.

Cyclosporine A, an immunosupressor drug already used in organ transplantation, and the addition of either hemoglobin or â-mercaptoethanol might inhibit mitochondrial-induced apoptosis in the ntESC. Inter-species incompatibility can be circumvented to a certain degree if the donor and host species are closely related. For instance, embryonic cells derived from the injection of a human nucleus in a chimpanzee ooplasm were viable, contrarily to when SCNT was done with the ooplasm of a nonhuman primate such as the orangutan.

The transfer of mitochondria isolated from patient-specific biopsies might circumvent the immune rejection problem due to mitochondrial heteroplasmy, and a female patient could in theory donate both the somatic nucleus and oocyte necessary for cell replacement therapy in her own body. The latter option offers great promises since recent studies in bovine showed that autologous SCNT embryonic production was more efficient than when the nuclear donor and ooplasmic host are from allogenic origins, and epigenetic reprogramming occurs to a significantly less extend in autologous SCNT embryos.

Transfer of Animal Contaminants

The interspecies transmission of pathogens is a nonnegligible issue when injecting a human nucleus into the oocyte of another species, such as bovine or pig. For instance, the porcine endogenous retrovirus (PERV), although inoffensive in pigs, disrupts the transcription initiation of genes in humans, as demonstrated in vitro by Moalic et al, by integrating within the CpG islands of promoters. Gene therapy approaches are being designed to reduce the infectivity of PERV due to the mannose-rich N-glycan integrated in the viral capsule. Thus, through the insertion of genes such as ManIb and ManII, encoding mannosidases involved in N-glycan catabolism, the infectivity of PERV was significantly reduced but not annihilated in human cells.

Animal serum and non-proliferative mouse fibroblast used as feeder-cell layer to direct the development of the human ntESC is problematic since animal contaminants can be transferred to the ntESC which in turn might trigger an immune response post-transplantation, thereby revoking the goal of therapeutic cloning.

N-glycolylneuraminic acid (Neu5Gc) is a mammalian sialic acid (a sugar with acidic side-chains present in the membrane of all cells) not found in humans, although most of us have anti-Neu5Gc antibodies. When human ntESC are grown on animal feeder cells or in contact with animal-derived serum, the stem cells incorporate enough Neu5Gc to potentially elicit an immune response in vivo, hence killing the transplanted cells. Although the murine leukemia virus is not pathogenic when transmitted from mouse feeder cells to human ntESC, non-cellular matrices are being designed to counteract the problem of animal contaminants.

Ludwig et al developed a combination of human growth factors (TGFβ, LiCl, bFGF, GABA, pipecolic acid which enhances receptor sensitivity to GABA) on a basal lamina reconstituted in vitro with human extracellular matrix components (collagen IV, laminin, vitronectin and fibronectin) as a substitute for an animal feeder-cell layer. Neu5Gc was not reported in the cell lines cultured on this human matrix. In sum, the issue of pathogenic transmission is in the process of being solved, bringing one step further the potential for clinical application of therapeutic cloning in cell replacement therapy.

Tumorgenesis and Spontaneous Differentiation

NtESC are subjected to the same tumorigenicity potential as wild-type stem cells. The formation of teratomas, after in vivo transplantation, is due to co-purification of pluripotent stem cells along with the wanted differentiated cells. Fujikawa et al noted tumor formation in the β-pancreatic cells transplanted in diabetic mice. The teratomas resulted from a low concentration of 0.2 per cent Oct4 and SSEA-1 positive cells, both markers of pluripotency downregulated upon differentiation, indicating the presence of left-over undifferentiated stem cells. Although the hyperglycemia associated with type I diabetes was reversed, tumorgenesis occurred 20 days post-transplantation, rendering stem cells, whether wild-type or issued from therapeutic cloning, a non-viable option for clinical applications in this instance, unless better isolation methods for the exclusive purification of differentiated stem cells are designed.

ETHICAL CONSIDERATIONS

Destruction of IVF

Therapeutic cloning and stem cell research stir an ethical controversy due to the source of embryonic stem cells, taken from aborted fetuses, unutilized zygotes and embryos morphologically incapable of in utero implantation, the latter representing 60 per cent of all embryos created through IVF. Some consider the use of discarded embryos diminutive since they imply "recycling" life products, as implied by Bush's statement that "there is no spare embryo."

The apparition of the primitive streak directing polarized development confers to the two-week embryo a higher moral status as a potential human organism, compared to the earlier embryo at the stage of a randomly-organized group of cells. Consequently, laws prohibiting the culture of embryos for more than two-weeks, which marks the onset of gastrulation and the formation of the primitive streak, are in vigor in several countries such as the United States, based on a decision of the British Warnock Commission in 1984.

The ethical debate on the moral impermissibility of deliberate destruction of an embryo can be circumvented by a new technique deviced by Chung et al. They successfully derived human ESC from a single cell without destroying the blastocyst in the process, using the same manipulations normally devoted to genetic screening in preimplantation embryos. This method seems to be promising for solving the ethical concern of killing a human embryo, rendering feasible the prenatal generation of individual-specific cell lines for use in regenerative medicine later on in life. However, Chung's method does not modify the fate of the ex vivo embryo, since the latter has a slim chance of implantation.

Moral Status of the IVF Embryo and the Argument of Potential

The main ethical roadblock against therapeutic cloning is the destruction of the generated embryos in order to collect cells that would further be differentiated in vitro. Embryo destruction is viewed as morally objectionable by the Prolife partisans because they grant the early embryo potential for personhood following development to term. Knowing that only 1 to 2 per cent of cloned mice produce viable organisms, the probability of producing a viable cloned human embryo is even slimmer.

However, the detractors of research using embryos would argue that the potential for personhood ought not to be granted on the basis of probability because a minority - one out of three -of zygotes conceived by natural means might implant in utero and be carried to term. Hence, the argument from potential relies on possibility rather than probability and is based on the belief that morally significant human life beings at conception. According to Dawson and Singer, "since something is logically impossible only if its assertion involves a contradiction, it is not logically impossible for a human blastocyst in a laboratory to develop into a person."

A counterargument put forth by the advocates of embryo research is that, left untouched, it is impossible that the in vitro embryo develops into the mature organism, and in utero implantation cannot occur if the transferred embryo reached the eight-celled stage and beyond. Nothing morally compelled one to generate a SCNT embryo that would have otherwise not existed, therefore we are not morally obligated to transfer the latter in utero. According to McMahan, "the idea that the potential to become a person confers a special moral status is plausible, if at all, only if the potential is identity-preserving.

"He further argues that, contrary to the late-term embryo, the early embryo as an "insentient cluster of cells" has nonidentity potential since the latter is not identical in morphology to the mature organism, which possesses a neuronal network characteristic of higher mental life, is able to develop complex cognitive capacities including self-consciousness and sentience, of which the early-term embryo is deprived. The destruction of an embryo of lower moral significance in the context of justified research to improve the quality of life of existing people of higher moral status ought to be viewed as morally permissible. Furthermore, under the philosophical point of view, "cell replacement therapy does not involve the destruction of an embryo but only its transformation into an embryonic cell line."

The engineering of mouse blastocysts lacking the Cdx2 gene, otherwise needed for the generation of the midgut endoderm and trophoblast differenciation, results in non-viable embryos that spontaneously stop dividing, providing a criticized alternative to the destruction of embryos by deliberate human action.

CURRENT LIMITATIONS OF CLONING TECHNOLOGY

Although promising, somatic cell nuclear transfer technology has certain limitations that require further improvements before therapeutic cloning can be applied widely in replacement therapy. Currently, the efficiency of the overall cloning process is low. The majority of embryos derived from animal cloning do not survive after implantation. In practical terms, multiple nuclear transfers must be performed to produce one live offspring for animal cloning applications.

The potential for cloned embryos to grow into live offspring is between 0.5 per cent to 18 per cent for sheep, cattle, pigs, and mice. However, greater success (80 per cent) has been reported in cattle, which may be in part due to the availability of advanced bovine supporting technologies, such as *in vitro* embryo production and embryo transfer, which have been developed for this species for agricultural purposes. To improve cloning efficiencies, further improvements are required in the multiple complex steps of nuclear transfer, such as enucleation and reconstruction, activation of oocytes, and cell cycle synchronization between donor cells and recipient oocytes, that will more readily produce viable sources of cells.

Furthermore, common abnormalities have been found in newborn clones if they survive to birth, including enlarged size with an enlarged placenta (large offspring syndrome), respiratory distress and defects of the kidney, liver, heart, and brain, obesity, and premature death. These may be related to the epigenetics of the cloned cells, which involve the reversible modifications of the DNA or chromatin, while the original DNA (genetic) sequences remain intact. Faulty epigenetic reprogramming in clones, where the DNA methylation patterns, histone modifications, and the overall chromatin structure of the somatic nuclei are not being reprogrammed to an embryonic pattern of expression, may explain the above abnormalities. Reactivation of key embryonic genes at the blastocyst stage is usually not present in embryos cloned from somatic cells, but embryos cloned from embryos consistently express early embryonic genes. Proper epigenetic reprogramming to an embryonic state may help to improve the cloning efficiency and reduce the incidence of abnormal cloned cells.

Tissue Engineering of Specific Structures

Investigators around the world, including our laboratory, have been working toward the development of several cell types and tissues and organs for clinical application.

Urethra

Various biomaterials without cells, such as PGA and acellular collagen-based matrices from small intestine and bladder, have been used experimentally (in animal models) for the regeneration of urethral tissue. Some of these biomaterials, like acellular collagen matrices derived from bladder submucosa, have also been seeded with autologous cells for urethral reconstruction.

Our laboratory has been able to replace tubularized urethral segments with cell-seeded collagen matrices. Acellular collagen matrices derived from bladder submucosa by our laboratory have been used experimentally and clinically. In animal studies, segments of the urethra were resected and replaced with acellular matrix grafts in an onlay fashion. Histologic examination showed complete epithelialization and progressive vessel and muscle infiltration, and the animals were able to void through the neourethras.

These results were confirmed in a clinical study of patients with hypospadias and urethral stricture disease. Decellularized cadaveric bladder submucosa was used as an onlay matrix for urethral repair in patients with stricture disease and hypospadias. Patent, functional neourethras were noted in these patients with up to a 7-yr follow-up. The use of an off-the-shelf matrix appears to be beneficial for patients with abnormal urethral conditions and obviates the need for obtaining autologous grafts, thus decreasing operative time and eliminating donor site morbidity. Unfortunately, the above techniques are not applicable for tubularized urethral repairs. The collagen matrices are able to replace urethral segments only when used in an onlay fashion. However, if a tubularized repair is needed, the collagen matrices should be seeded with autologous cells to avoid the risk of stricture formation and poor tissue development. Therefore, tubularized collagen matrices seeded with autologous cells can be used successfully for total penile urethra replacement.

Bladder

Currently, gastrointestinal segments are commonly used as tissues for bladder replacement or repair. However, gastrointestinal tissues are designed to absorb specific solutes, whereas bladder tissue is designed for the excretion of solutes. Because of the problems encountered with the use of gastrointestinal segments, numerous investigators have attempted alternative materials and tissues for bladder replacement or repair. The success of the use of cell transplantation strategies for bladder reconstruction depends on the ability to use donor tissue efficiently and to provide the right conditions for long term survival, differentiation and growth.

Urothelial and muscle cells can be expanded *in vitro*, seeded onto polymer scaffolds, and allowed to attach and form sheets of cells. These principles were applied toward the creation of tissue engineered bladders in an animal model that required a subtotal cystectomy with subsequent replacement with a tissue engineered organ in beagle dogs. Urothelial and muscle cells were separately expanded from an autologous bladder biopsy sample and seeded onto a bladder-shaped biodegradable polymer scaffold. The results from this study showed that it is possible to tissue-engineer bladders that are anatomically and functionally normal.

Male Genital Tissues

Reconstructive surgery is required for a wide variety of pathologic penile conditions, such as penile carcinoma, trauma, severe erectile dysfunction, and congenital conditions such as ambiguous genitalia, hypospadias, and epispadias. One of the major limitations of phallic reconstructive surgery is the scarcity of sufficient autologous tissue. Phallic reconstruction using autologous tissue, derived from the patient's own cells, may be preferable in selected cases. The major components of the phallus are corporal smooth muscle and endothelial cells. The creation of autologous functional and structural corporal tissue *de novo* would be beneficial.

Autologous cavernosal smooth muscle and endothelial cells were harvested, expanded and seeded on acellular collagen matrices and implanted in a rabbit model. Histologic examination confirmed the appropriate organization of penile tissue phenotypes, and structural and functional studies, including cavernosonography, cavernosometry, and mating studies, demonstrated that it is possible to engineer autologous functional penile tissue. Our laboratory is currently working on increasing the size of the engineered constructs.

Female Genital Tissues

Congenital malformations of the uterus may have profound implications clinically. Patients with cloacal exstrophy and intersex disorders may not have sufficient uterine tissue present for future reproduction. We investigated the possibility of engineering functional uterine tissue using autologous cells. Autologous rabbit uterine smooth muscle and epithelial cells were harvested, then grown and expanded in culture. These cells were seeded onto preconfigured uterine-shaped biodegradable polymer scaffolds, which were then used for subtotal uterine tissue replacement in the corresponding autologous animals. Upon retrieval 6 mo after implantation, histologic, immunocytochemical, and Western blot analyses confirmed the presence of normal uterine tissue components. Biomechanical analyses and organ bath studies showed that the functional characteristics of these tissues were similar to those of normal uterine tissue.

Breeding studies that use these engineered uteri are currently being performed. Similarly, several pathologic conditions, including congenital malformations and malignancy, can adversely affect normal vaginal development or anatomy. Vaginal reconstruction has traditionally been challenging because of the paucity of available native tissue. The feasibility of engineering vaginal tissue*in vivo* was investigated.

Vaginal epithelial and smooth muscle cells of female rabbits were harvested, grown, and expanded in culture. These cells were seeded onto biodegradable polymer scaffolds, and the cell-seeded constructs were then implanted into nude mice for up to 6 wk. Immunocytochemical, histologic, and Western blot analyses confirmed the presence of vaginal tissue phenotypes. Electrical field stimulation studies in the tissue-engineered constructs showed similar functional properties to those of normal vaginal tissue. When these constructs were used for autologous total vaginal replacement, patent vaginal structures were noted in the tissue-engineered specimens, whereas the non–cell-seeded structures were noted to be stenotic.

Kidney

We applied the principles of both tissue engineering and therapeutic cloning in an effort to produce genetically identical renal tissue in a large animal model, cattle. Bovine skin fibroblasts from adult Holstein steers were obtained by ear notch, and single donor cells were isolated and microinjected into the perivitelline space of donor enucleated oocytes (nuclear transfer). The resulting blastocysts were implanted into progastrin-synchronized recipients to allow for further *in vivo* growth. After 12 wk, cloned renal cells were harvested, expanded *in vitro,* and seeded onto biodegradable scaffolds. The constructs, which consisted of the cells and the scaffolds, were then implanted into the subcutaneous space of the same steer from which the cells were cloned to allow for tissue growth.

The kidney is a complex organ with multiple cell types and a complex functional anatomy that renders it one of the most difficult to reconstruct. Previous efforts in tissue engineering of the kidney have been directed toward the development of extracorporeal renal support systems made of biologic and synthetic components, and *ex vivo*renal replacement devices are known to be life-sustaining. However, there would be obvious benefits for patients with end-stage kidney disease if these devices could be implanted long term without the need for an extracorporeal perfusion circuit or immunosuppressive drugs.

Cloned renal cells were seeded on scaffolds consisting of three collagen-coated cylindrical polycarbonate membranes. The ends of the three membranes of each scaffold were connected to catheters that terminated into a collecting reservoir. This created a renal neo-organ with a mechanism for collecting the excreted urinary fluid. These scaffolds with the collecting devices were transplanted subcutaneously into the same steer from which the genetic material originated and retrieved 12 wk after implantation. Chemical analysis of the collected urinelike fluid, including urea nitrogen and creatinine levels, electrolyte levels, specific gravity, and glucose concentration, revealed that the implanted renal cells possessed filtration, reabsorption, and secretory capabilities.

Histologic examination of the retrieved implants revealed extensive vascularization and self-organization of the cells into glomeruli- and tubulelike structures. A clear continuity between the glomeruli, the tubules, and the polycarbonate membrane was noted that allowed the passage of urine into the collecting reservoir. Immunohistochemical analysis with renal-specific antibodies revealed the presence of renal proteins, RT-PCR analysis confirmed the transcription of renal specific RNA in the cloned specimens, and Western blot analysis confirmed the presence of elevated renal-specific protein levels.

Because previous studies have shown that bovine clones harbor the oocyte mtDNA, the donor egg's mtDNA was thought to be a potential source of immunologic incompatibility. Differences in mtDNA-encoded proteins expressed by cloned cells could stimulate a T cell response specific for mtDNA-encoded minor histocompatibility antigens when the cloned cells are implanted back into the original nuclear donor. We used nucleotide sequencing of the mtDNA genomes of the clone and fibroblast nuclear donor to identify potential antigens in the muscle constructs. Only two amino acid substitutions were noted to distinguish the clone and the nuclear donor, and as a result, a maximum of two minor histocompatibility antigens could be defined. Given the lack of knowledge regarding peptide-binding motifs for bovine MHC class I molecules, there is no reliable method to predict the effect of these amino acid substitutions on bovine histocompatibility.

Oocyte-derived mtDNA was also thought to be a potential source of immunologic incompatibility in the cloned renal cells. Maternally transmitted minor histocompatibility antigens in mice have been shown to stimulate both skin allograft rejection *in vivo* and cytotoxic T lymphocytes expansion *in vitro,* which could prevent the use of these cloned constructs in patients with chronic rejection of major histocompatibility matched human renal transplants. We tested for a possible T cell response to the cloned renal devices using delayed-type hypersensitivity testing *in vivo* and ELISPOT analysis of IFN-^{3}–secreting T cells *in vitro*. Both analyses revealed that the cloned renal cells showed no evidence of a T cell response, suggesting that rejection will not necessarily occur in the presence of oocyte-derived mitochondrial DNA. This finding may represent a step forward in overcoming the histocompatibility problem of stem cell therapy.

These studies demonstrated that cells derived from nuclear transfer can be successfully harvested, expanded in culture, and transplanted *in vivo* with the use of biodegradable scaffolds on which the single suspended cells can organize into tissue structures that are genetically identical to that of the host. These studies were the first demonstration of the use of therapeutic cloning for regeneration of tissues *in vivo*.

Blood Vessels

Xenogenic or synthetic materials have been used as replacement blood vessels for complex cardiovascular lesions. However, these materials typically lack growth potential and may place the recipient at risk for complications such as stenosis, thromboembolization, or infection.

Tissue-engineered vascular grafts have been constructed using autologous cells and biodegradable scaffolds and have been applied in dog and lamb models. The key advantage from the use these autografts is that they degrade *in vivo* and thus allow the new tissue to form without the long-term presence of foreign material. Application of these techniques from the laboratory to the clinical setting have begun, where autologous vascular cells were harvested, expanded, and seeded onto a biodegradable scaffold. The resultant autologous construct was used to replace a stenosed pulmonary artery that had been previously repaired. Seven months after implantation, no evidence of graft occlusion or aneurysmal changes were noted in the recipient.

Cartilage—Articular Cartilage and Trachea

Full-thickness articular cartilage lesions have limited healing capacity and thus represent a difficult management issue for the clinicians who treat adult patients with damaged articular cartilage. Large defects can be associated with mechanical instability and may lead to degenerative joint disease if left untreated. Chondrocytes were expanded and cultured onto biodegradable scaffolds to create engineered cartilage for use in large osteochondral defects in rabbits.

When sutured to a subchondral support, the engineered cartilage was able to withstand physiologic loading and underwent orderly remodeling of the large osteochondral defects in adult rabbits. Thus, the engineered cartilage was able to provide a biomechanically functional template that was able to undergo orderly remodeling when subjected to quantitative structural and functional analyses.

Few treatment options are currently available for patients who have severe congenital tracheal pathology, such as stenosis, atresia, and agenesis, because of the limited availability of autologous transplantable tissue in the neonatal period. Tissue engineering in the fetal period may be a viable alternative for the surgical treatment of these prenatally diagnosed congential anomalies because cells could be harvested and grown into transplantable tissue in parallel with the remainder of gestation.

Chondrocytes from both elastic and hyaline cartilage specimens have been harvested from fetal lambs, expanded *in vitro,* then dynamically seeded onto biodegradable scaffolds. The constructs were then implanted as replacement tracheal tissue in fetal lambs. The resultant tissue-engineered cartilage was noted to undergo engraftment and epithelialization while maintaining its structural support and patency. Furthermore, if native tracheal tissue is unavailable, engineered cartilage may be derived from bone marrow–derived mesenchymal progenitor cells as well.

CELLULAR THERAPIES

Bulking Agents

Injectable bulking agents can be endoscopically used in the treatment of both urinary incontinence and vesicoureteral reflux. The advantages in treating urinary incontinence and vesicoureteral reflux with this minimally invasive approach include the simplicity of this quick outpatient procedure and the low morbidity associated with it. Several investigators are seeking alternative implant materials that would be safe for human use. The ideal substance for the endoscopic treatment of reflux and incontinence should be injectable, nonantigenic, nonmigratory, volume stable, and safe for human use. Toward this goal, long-term studies were conducted to determine the effect of injectable chondrocytes *in vivo.*

It was initially determined that alginate, a liquid solution of gluronic and mannuronic acid, embedded with chondrocytes, could serve as a synthetic substrate for the injectable delivery and maintenance of cartilage architecture *in vivo*. Alginate undergoes hydrolytic biodegradation and its degradation time can be varied depending on the concentration of each of the polysaccharides. The use of autologous cartilage for the treatment of vesicoureteral reflux in humans would satisfy all of the requirements for an ideal injectable substance.

Chondrocytes derived from an ear biopsy can be readily grown and expanded in culture. Neocartilage formation can be achieved *in vitro* and *in vivo* by using chondrocytes cultured on synthetic biodegradable polymers. In these experiments, the cartilage matrix replaced the alginate as the polysaccharide polymer underwent biodegradation. This system was adapted for the treatment of vesicoureteral reflux in a porcine model. These studies showed that chondrocytes can be easily harvested and combined with alginate *in vitro*, the suspension can be easily injected cystoscopically, and the elastic cartilage tissue formed is able to correct vesicoureteral reflux without any evidence of obstruction. Two multicenter clinical trials were conducted by using the above engineered chondrocyte technology. Patients with vesicoureteral reflux were treated at 10 centers throughout the United States.

The patients had a similar success rate as with other injectable substances in terms of cure. Chondrocyte formation was not noted in patients who experienced treatment failure. The patients who were cured would supposedly have a biocompatible region of engineered autologous tissue present, rather than a foreign material. Patients with urinary incontinence were also treated endoscopically with injected chondrocytes at three different medical centers. Phase 1 trials showed an approximate success rate of 80 per cent at both 3 and 12 mo postoperatively.

Injectable Muscle Cells

The potential use of injectable, cultured myoblasts for the treatment of stress urinary incontinence has been investigated. Labeled myoblasts were directly injected into the proximal urethra and lateral bladder walls of nude mice with a microsyringe in an open surgical procedure. Tissue harvested up to 35d after injection contained the labeled myoblasts, as well as evidence of differentiation of the labeled myoblasts into regenerative myofibers. The authors reported that a significant portion of the injected myoblast population persisted *in vivo*. Similar techniques of sphincteric derived muscle cells have been used for the treatment of urinary incontinence in a pig model. The fact that myoblasts can be labeled and survive after injection and begin the process of myogenic differentiation further supports the feasibility of the use of cultured cells of muscular origin as an injectable bioimplant.

The use of injectable muscle precursor cells has also been investigated for use in the treatment of urinary incontinence due to irreversible urethral sphincter injury or maldevelopment. Muscle precursor cells are the quiescent satellite cells found in each myofiber that proliferate to form myoblasts and eventually myotubes and new muscle tissue. Intrinsic muscle precursor cells have previously been shown to play an active role in the regeneration of injured striated urethral sphincter.

In a subsequent study, autologous muscle precursor cells were injected into a rat model of urethral sphincter injury, and both replacement of mature myotubes as well as restoration of functional motor units were noted in the regenerating sphincteric muscle tissue. This is the first demonstration of the replacement of both sphincter muscle tissue and its innervation by the injection of muscle precursor cells. As a result, muscle precursor cells may be a minimally invasive solution for urinary incontinence in patients with irreversible urinary sphincter muscle insufficiency.

Endocrine Replacement

Patients with testicular dysfunction and hypogonadal disorders are dependent on androgen replacement therapy to restore and maintain physiologic levels of serum testosterone and its metabolites, dihydrotestosterone and estradiol. Currently available androgen replacement modalities, such as testosterone tablets and capsules, depot injections, and skin patches, may be associated with fluctuating serum levels and complications such as fluid and nitrogen retention, erythropoiesis, hypertension, and bone density changes. Because Leydig cells of the testes are the major source of testosterone in men, implantation of heterologous Leydig cells or gonadal tissue fragments has previously been proposed as a method for chronic testosterone replacement. But these approaches were limited by the failure of the tissues and cells to produce testosterone.

Encapsulation of cells in biocompatible and semipermeable polymeric membranes has been an effective method to protect against a host immune response as well as to maintain viability of the cells while allowing the secretion of desired therapeutic agents. Alginate-poly-L-lysine–encapsulated Leydig cell microspheres were used as a novel method for testosterone delivery *in vivo*. Elevated stable serum testosterone levels were noted in castrated adult rats over the course of the study, suggesting that microencapsulated Leydig cells may be a potential therapeutic modality for testosterone supplementation.

Angiogenic Agents

The engineering of large organs will require a vascular network of arteries, veins, and capillaries to deliver nutrients to each cell. One possible method of vascularization is through the use of gene delivery of angiogenic agents such as vascular endothelial growth factor (VEGF) with the implantation of vascular endothelial cells to enhance neovascularization of engineered tissues. Skeletal myoblasts from adult mice were cultured and transfected with an adenovirus encoding VEGF and combined with human vascular endothelial cells. The mixtures of cells were injected subcutaneously in nude mice, and the engineered tissues were retrieved up to 8wk after implantation. The transfected cells were noted to form muscle with neovascularization by histology and immunohistochemical probing with maintenance of their muscle volume, whereas engineered muscle of nontransfected cells had a significantly smaller mass of cells with loss of muscle volume over time, less neovascularization, and no surviving endothelial cells. These results indicate that a combination of VEGF and endothelial cells may be useful for inducing neovascularization and volume preservation in engineered tissue.

Antiangiogenic Agents

The delivery of antiangiogenic agents may help to slow tumor growth for a variety of neoplasms. Encapsulated hamster kidney cells transfected with the angiogenesis inhibitor endostatin were used for local delivery on human glioma cell line xenografts. The release of biologically active endostatin led to significant inhibition of endothelial cell proliferation and substantial reduction in tumor weight. Continuous local delivery of endostatin via encapsulated endostatin-secreting cells may be effective therapeutic option for a variety of tumor types.

Tissue engineering efforts are currently underway for virtually every type of tissue and organ within the human body. Because tissue engineering incorporates the fields of cell transplantation, materials science, and engineering, personnel who have mastered the techniques of cell harvest, culture, expansion, transplantation, and polymer design are essential for the successful application of this technology. Various engineered tissues are at different stages of development, with some already being used clinically, a few in preclinical trials, and some in the discovery stage. Recent progress suggests that engineered tissues may have an expanded clinical applicability in the future because they represent a viable therapeutic option for those who require tissue replacement. More recently, major advances in the areas of stem cell biology, tissue engineering, and nuclear transfer techniques have made it possible to combine these technologies to create the comprehensive scientific field of regenerative medicine.

3

Plant Cell and Tissue Culture Technology

Most methods of plant transformation applied to GM crops require that a whole plant is regenerated from isolated plant cells or tissue which have been genetically transformed. This regeneration is conducted in vitro so that the environment and growth medium can be manipulated to ensure a high frequency of regeneration. In addition to a high frequency of regeneration, the regenerable cells must be accessible to gene transfer by whatever technique is chosen.

The primary aim is therefore to produce, as easily and as quickly as possible, a large number of regenerable cells that are accessible to gene transfer. The subsequent regeneration step is often the most difficult step in plant transformation studies. However, it is important to remember that a high frequency of regeneration does not necessarily correlate with high transformation efficiency. It will also look at the basic culture types used for plant transformation and cover some of the techniques that can be used to regenerate whole transformed plants from transformed cells or tissue. Practically any plant transformation experiment relies at some point on tissue culture. There are some exceptions to this generalisation, but the ability to regenerate plants from isolated cells or tissues in vitro underpins most plant transformation systems.

PLANT CELL

Plant cells are eukaryotic cells, or cells with a membrane-bound nucleus. Unlike prokaryotic cells, the DNA in a plant cell is housed within the nucleus. In addition to having a nucleus, plant cells also contain other membrane-bound organelles, or tiny cellular structures, that carry out specific functions necessary for normal cellular operation.

Organelles have a wide range of responsibilities that include everything from producing hormones and enzymes to providing energy for a plant cell. Plant cells are similar to animal cells in that they are both eukaryotic cells and have similar organelles. Plant cells are generally larger than animal cells. While animal cells come in various sizes and tend to have irregular shapes, plant cells are more similar in size and are typically rectangular or cube shaped. A plant cell also contains structures not found in an animal cell. Some of these include a cell wall, a large vacuole, and plastids. Plastids, such as chloroplasts, assist in storing and harvesting needed substances for the plant. Animal cells also contain structures such as centrioles, lysosomes, and cilia and flagella that are not typically found in plant cells.

GENERAL PLANT ORGANISATION

A plant has two organ systems:

1. The shoot system, and
2. The root system.

The shoot system is above ground and includes the organs such as leaves, buds, stems, flowers (if the plant has any), and fruits (if the plant has any). The root system includes those parts of the plant below ground, such as the roots, tubers, and rhizomes.

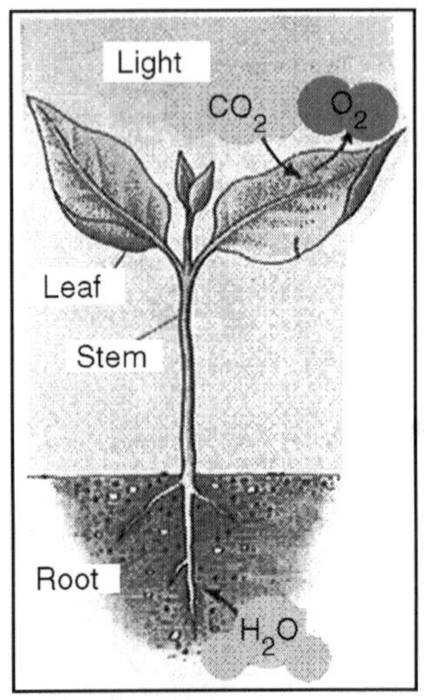

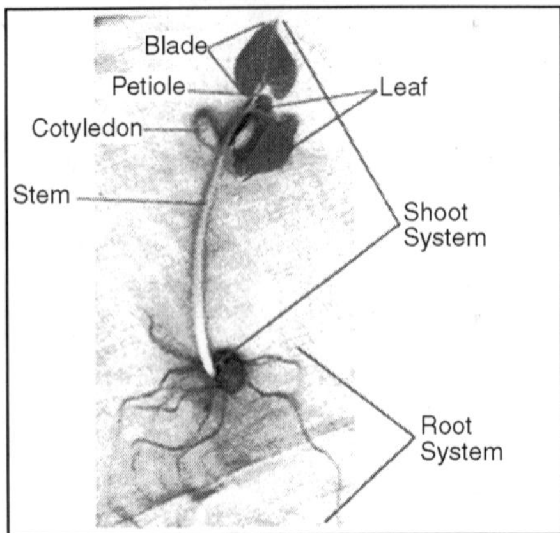

Fig. Major Organ Systems of the Plant Body.

Plant cells are formed at meristems, and then develop into cell types which are grouped into tissues.

Plants have only three tissue types:

1. Dermal;
2. Ground; and
3. Vascular.

Dermal tissue covers the outer surface of herbaceous plants. Dermal tissue is composed of epidermal cells, closely packed cells that secrete a waxy cuticle that aids in the prevention of water loss. The ground tissue comprises the bulk of the primary plant body. Parenchyma, collenchyma, and sclerenchyma cells are common in the ground tissue. Vascular tissue transports food, water, hormones and minerals within the plant. Vascular tissue includes xylem, phloem, parenchyma, and cambium cells.

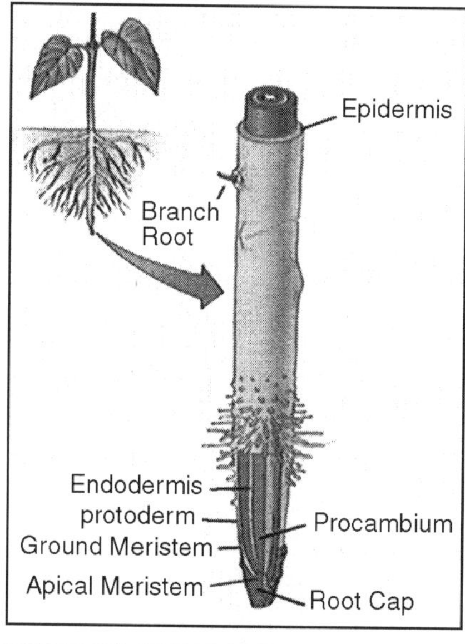

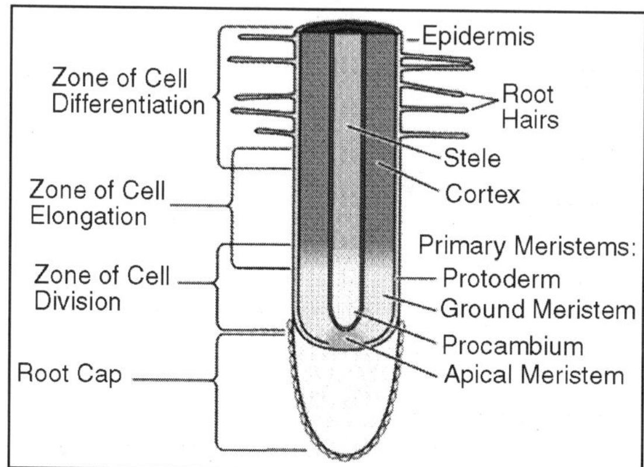

Fig. Two Views of the Structure of the Root and Root Meristem.

Plant cell types rise by mitosis from a meristem. A meristem may be defined as a region of localised mitosis. Meristems may be at the tip of the shoot or root (a type known as theapical meristem) or lateral, occurring in cylinders extending nearly the length of the plant. A cambium is a lateral meristem that produces (usually) secondary growth. Secondary growth produces both wood and cork (although from separate secondary meristems).

Parenchyma

A generalised plant cell type, parenchyma cells are alive at maturity. They function in storage, photosynthesis, and as the bulk of ground and vascular tissues. Palisade parenchyma cells are elogated cells located in many leaves just below the epidermal tissue.

Spongy mesophyll cells occur below the one or two layers of palisade cells. Ray parenchyma cells occur in wood rays, the structures that transport materials laterally within a woody stem. Parenchyma cells also occur within the xylem and phloem of vascular bundles. The largest parenchyma cells occur in the pith region, often, as in corn (Zea) stems, being larger than the vascular bundles. In many prepared slides they stain green.

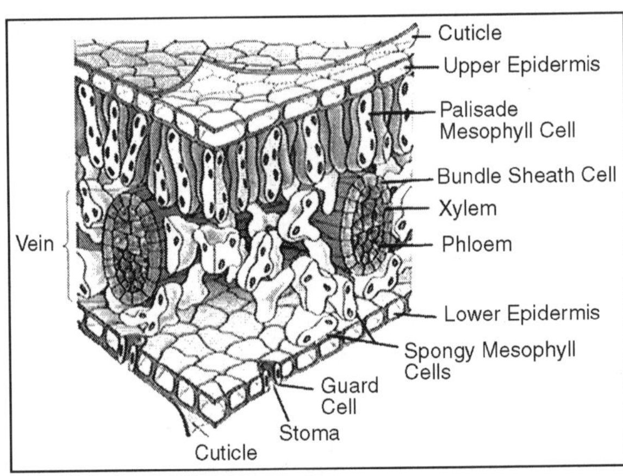

Fig. Diagram of Leaf Structure.

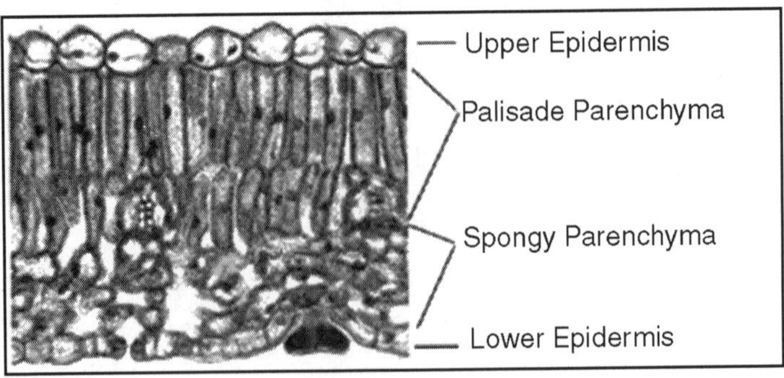

Fig. Cross-section of a Stained Leaf of *Syringia*.

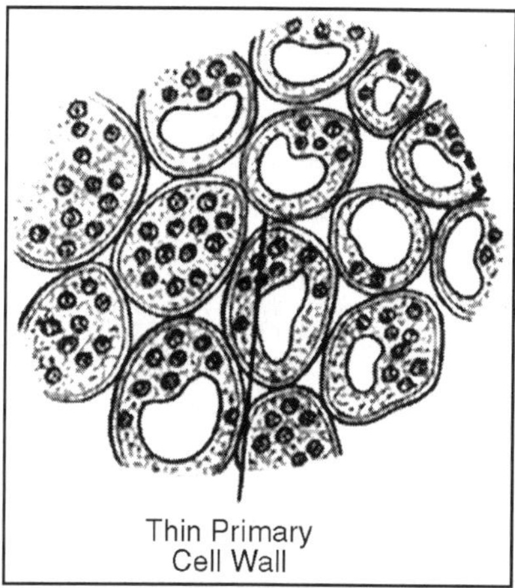

Fig. Diagram of a Series of Parenchyma Cells.

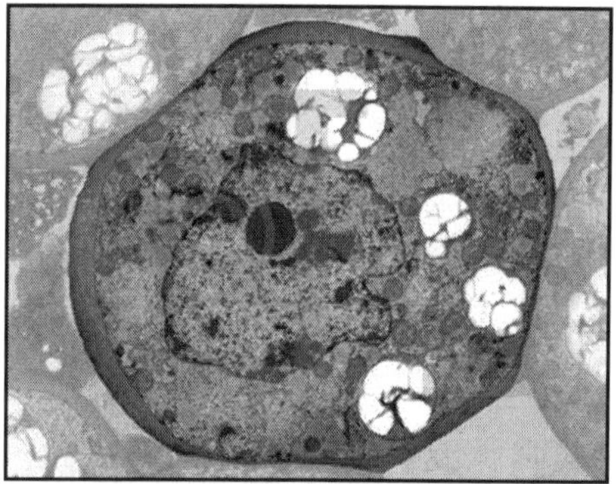

Fig. Lily Parenchyma Cell (cross-section).

Collenchyma

Collenchyma cells support the plant. These cells are characterised by thickenings of the wall, the are alive at maturity. They tend to occur as part of vascular bundles or on the corners of angular stems. In many prepared slides they stain red.

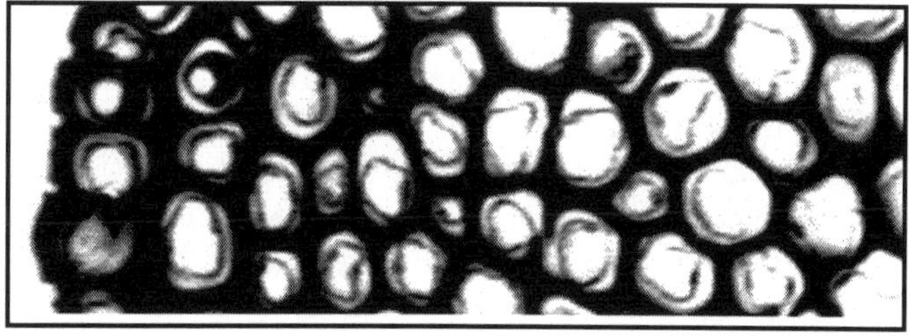

Fig. Collenchyma Cells. Note the Thick Walls on the Collenchyma Cells

Occurring at the Edges of the *Medicago* Stem Cross Section.

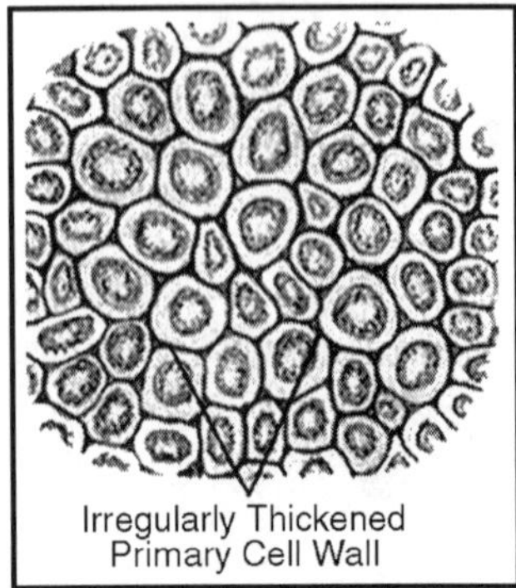

Fig. Diagram of Some Collenchyma Cells.

Sclerenchyma

Sclerenchyma cells support the plant. They often occur as bundle cap fibres. Sclerenchyma cells are characterised by thickenings in their secondary walls. They are dead at maturity. They, like collenchyma, stain red in many commonly used prepared slides. A common type of schlerenchyma cell is the fibre.

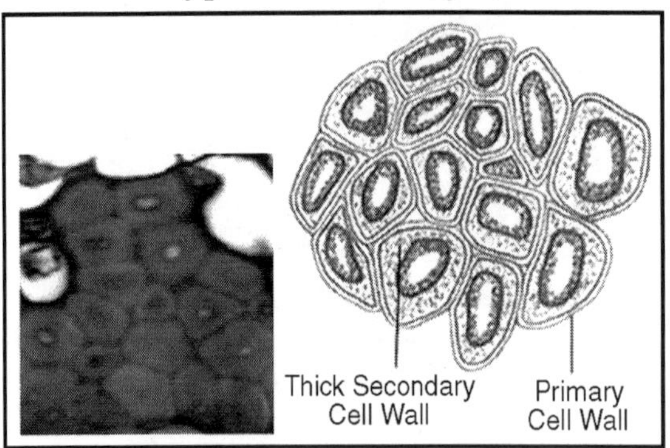

Fig. Sclerenchyma Cells.

Xylem

Xylem is a term applied to woody (lignin-impregnated) walls of certain cells of plants. Xylem cells tend to conduct water and minerals from roots to leaves. While parenchyma cells do occur within what is commonly termed the "xylem" the more identifiable cells, tracheids and vessel elements, tend to stain red with Safranin-O. Tracheids are the more primitive of the two cell types, occurring in the earliest vascular plants. Tracheids are long and tapered, with angled end-plates that connect cell to cell. Vessel elements are shorter, much wider, and lack end plates. They occur only in angiosperms, the most recently evolved large group of plants.

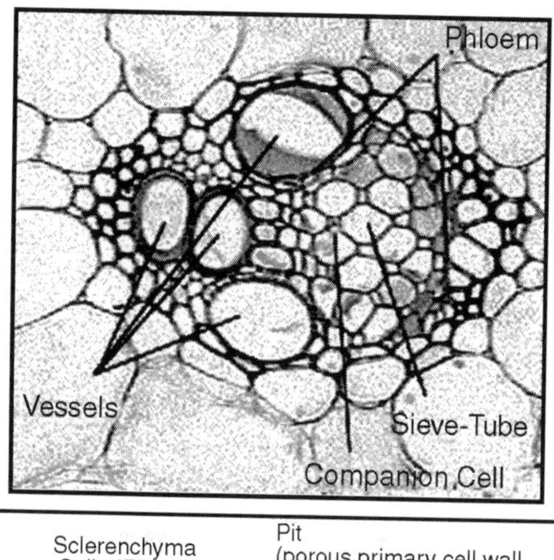

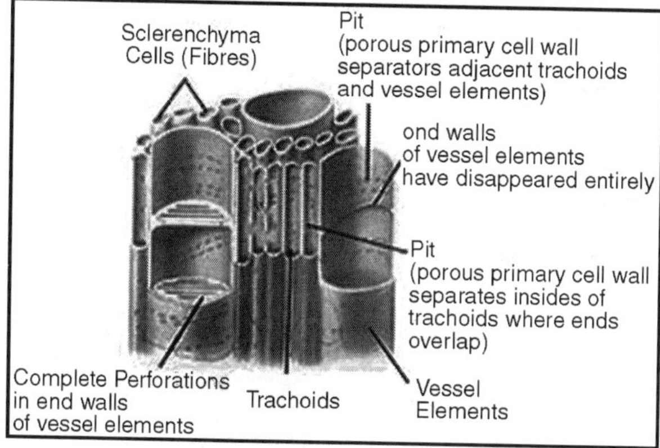

Fig. Xylem Cells.

Tracheids, longer, and narrower than most vessels, appear first in the fossil record. Vessels occur later. Tracheids have obliquely-angled endwalls cut across by bars. The evolutionary trend in vessels is for shorter cells, with no bars on the endwalls.

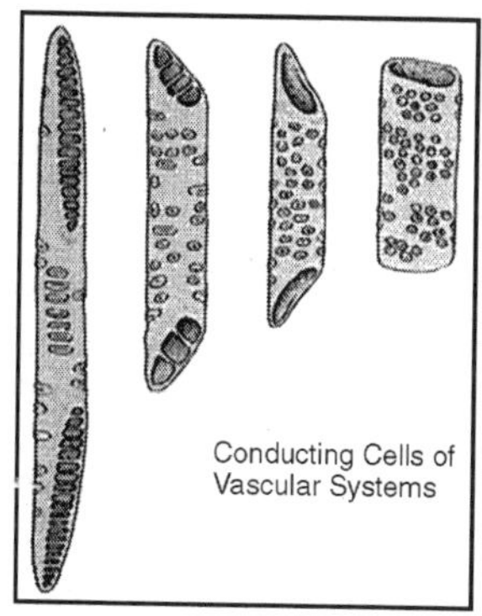

Fig. Conducting Cells of the Xylem; Tracheids (Left) are More Primitive, While the Various Types of Vessels (the other Three) are more Advanced.

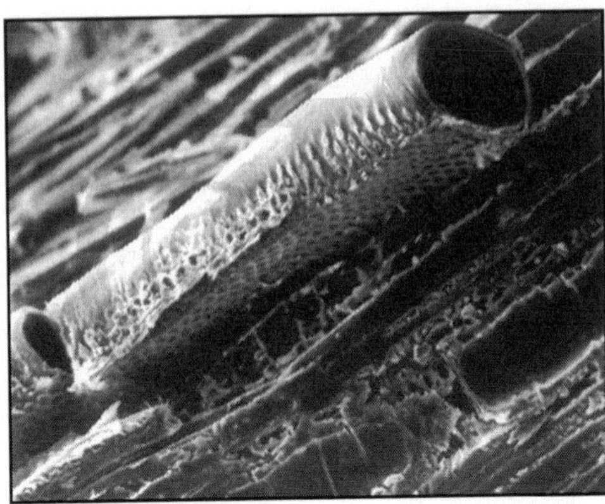

Fig. Conductive Vessel Element in Mountain Mahogany Wood.

Phloem cells conduct food from leaves to rest of the plant. They are alive at maturity and tend to stain green (with the stain fast green). Phloem cells are usually located outside the xylem. The two most common cells in the phloem are the companion cells and sieve cells. Companion cells retain their nucleus and control the adjacent sieve cells. Dissolved food, as sucrose, flows through the sieve cells.

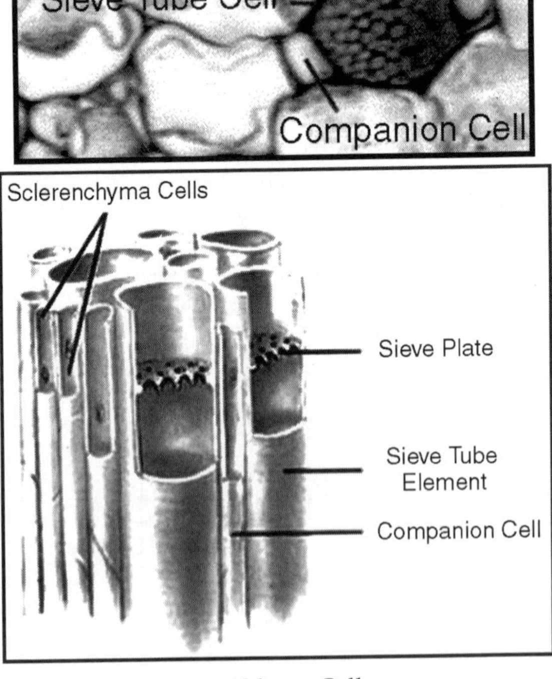

Fig. Phloem Cells.

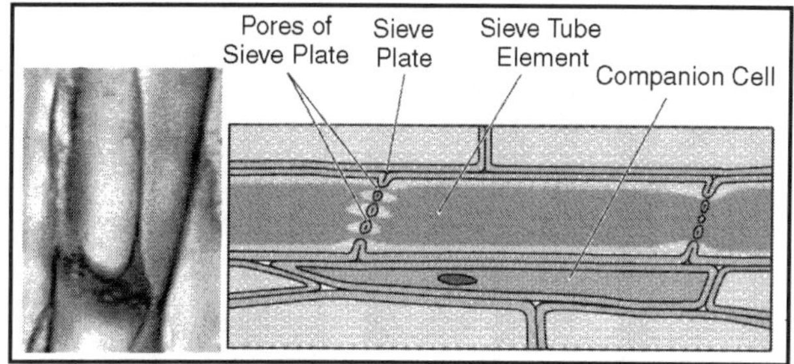

Fig. Phloem Cells as seen in Longitudinal Section.

EPIDERMAL CELLS

Epidermis

The epidermal tissue functions in prevention of water loss and acts as a barrier to fungi and other invaders. Thus, epidermal cells are closely packed, with little intercellular space. To further cut down on water loss, many plants have a waxy cuticle layer deposited on top of the epidermal cells.

Guard Cells

To facilitate gas exchange between the inner parts of leaves, stems, and fruits, plants have a series of openings known as stomata (singular stoma). Obviously these openings would allow gas exchange, but at a cost of water loss. Guard cells are bean-shaped cells covering the stomata opening. They regulate exchange of water vapour, oxygen and carbon dioxide through the stoma.

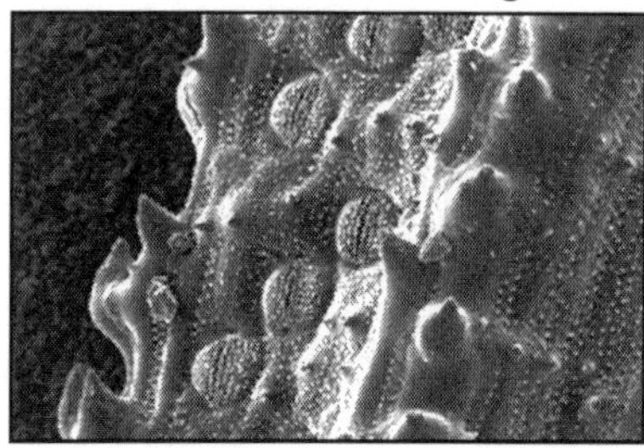

Fig. Scanning Electron Micrograph of *Equisetum* (Horsetail or Scouring Rush) Epidermis.

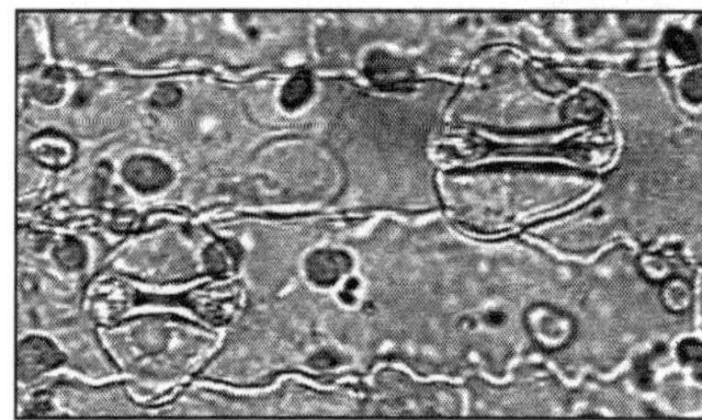

Fig. Epidermal Cells, Including Guard Cells, of Corn.

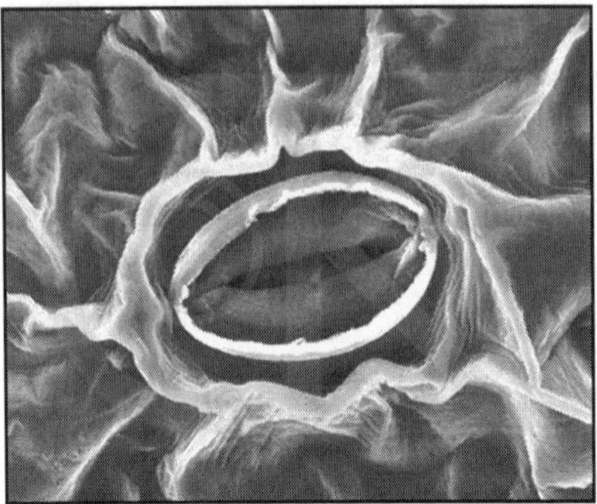

Fig. Pea Leaf Stoma, *Vicea* sp.

Monocots and Dicots

Angiosperms, flowering plants, are divided into two groups: monocots and dicots.

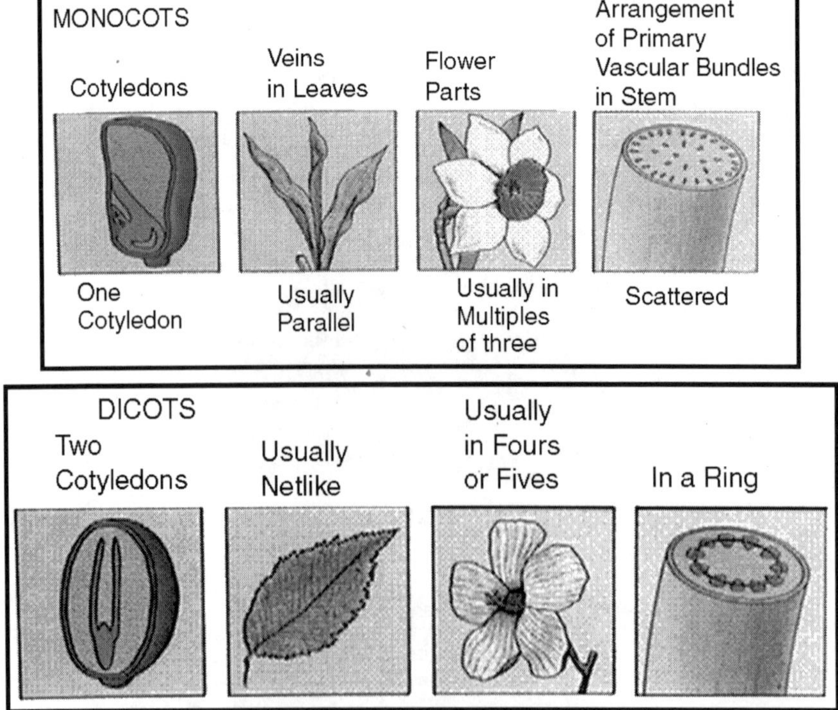

Fig. Features of Monocot and Dicot Plants.

Monocot seeds have one "seed leaf" termed a cotyledon (in fact monocot is a shortening of monocotyledon). Dicots have two cotyledons. Both groups, however, have the same basic architecture of nodes, internodes, etc.

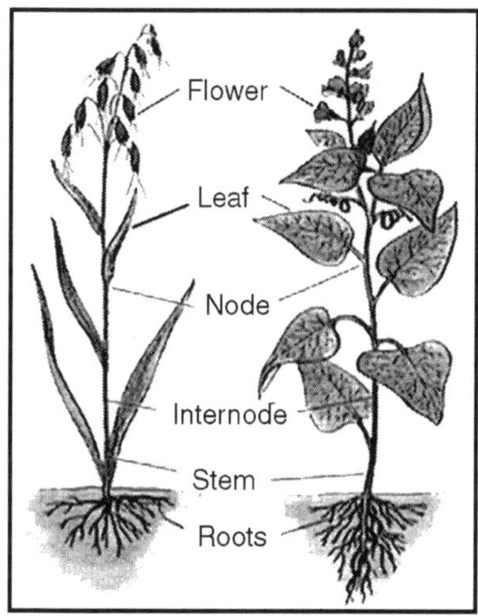

Fig. Comparison of Monocot (Left, Oat) and Dicot (Right, Bean) Gross Anatomy.

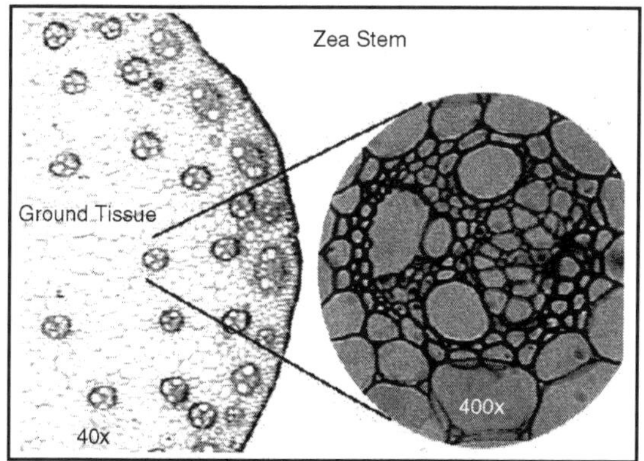

Fig. The Scattered Vascular Bundles of the Corn Stem.

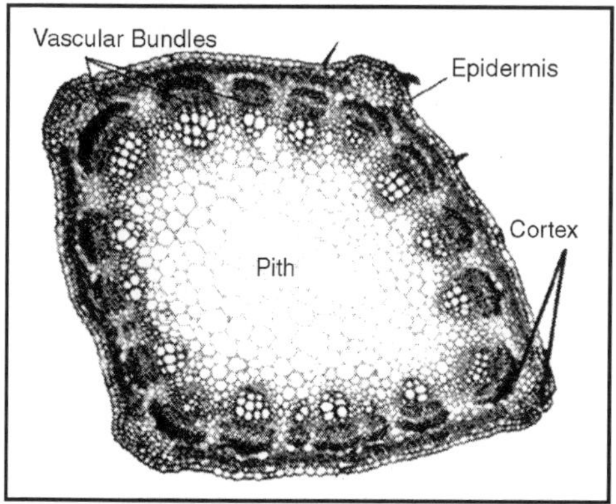

Fig. The Ringed Array of Vascular Bundles in this Dicot Stem (*Medicago*).

Monocot stems have scattered vascular bundles. Dicot stems have their vascular bundles in a ring arrangement. Monocot stems have most of their vascular bundles near the outside edge of the stem. The bundles are surrounded by large parenchyma in the cortex region. There is no pith region in monocots.

Dicot stems have bundles in a ring surrounding parenchyma cells in a pith region. Between the bundles and the epidermis are smaller (as compared to the pith) parenchyma cells making up the cortex region.

Monocot roots, interestingly, have their vascular bundles arranged in a ring. Dicot roots have their xylem in the centre of the root and phloem outside the xylem. A carrot is an example of a dicot root.

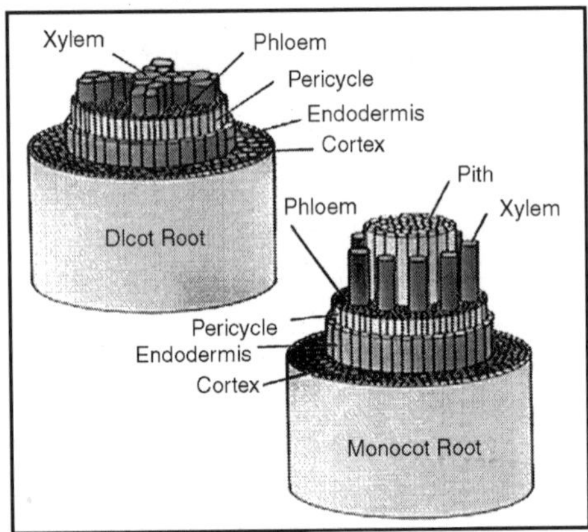

Fig. Diagram Illustrating the Tissue Layers and their Organisation Within monocot and Dicot Roots.

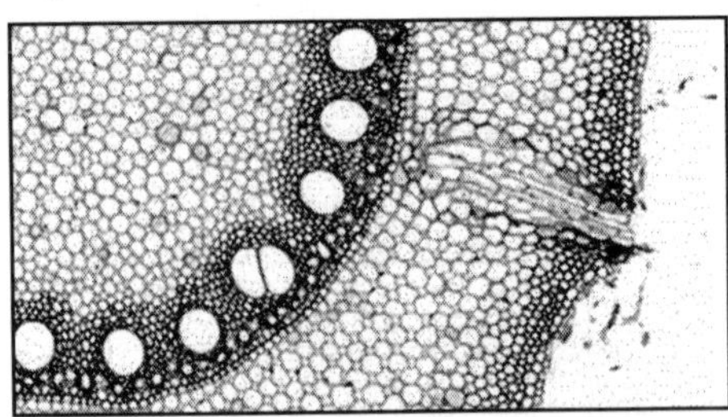

Fig. Cross-section of a Root of Corn.

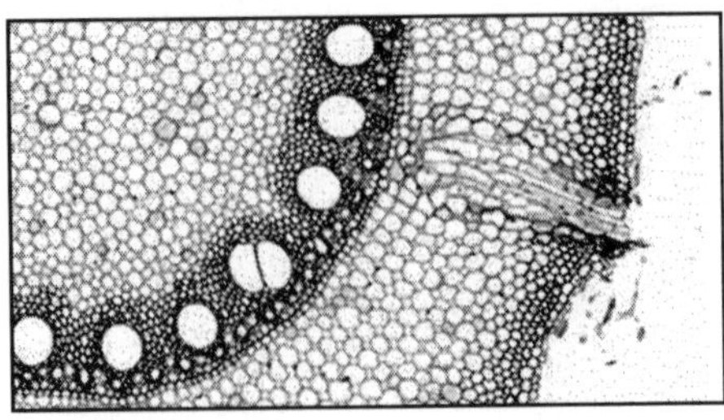

Fig. Cross-section of a Dicot Root.

Monocot leaves have their leaf veins arranged parallel to each other and the long axis of the leaf (parallel vennation). An common example of this is the husk of corn or a blade of grass (both are monocots). Dicot leaves have an anastamosing network of veins arising from a mid-vein termed net vennation. Examples of dicot leaves include maples, oaks, geraniums, and dandelions.

Monocots have their flower parts in threes or multiples of three; example the tulip and lily (*Lilium*). Dicots have their flower parts in fours (or multiples) or fives (or multiples). Examples of some common dicot flowers include the geranium, snapdragon, and citrus.

Fig. Monocot (Left) and Dicot (Right) Flowers.

Secondary Growth

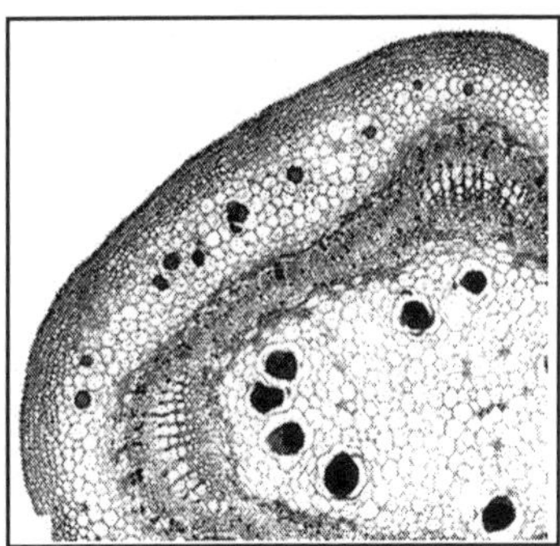

Fig. Cross-section of a Young Stem of Basswood.

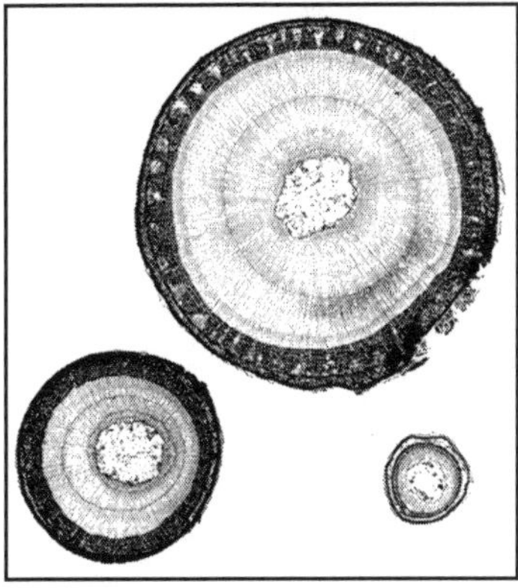

Fig. Three Cross-sections of Older Basswood Twigs.

Secondary growth is produced by a cambium. It occurs in rows or ranks of cork, secondary xylem or secondary phloem cells. Cork cells (produced by a cork cambium) are technically part of the epidermis, and contribute to the bark of woody stems. Dicot secondary growth occurs by growth of vascular cambium, to complete a full vascular cylinder around the plant. Secondary xylem is produced to the inside of the vascular cambium, secondary phloem to the outside. The living parts of the woody plant are next to the vascular cambium.

At the end of each growing season, the vascular cambium stops growing, forming a growth ring.

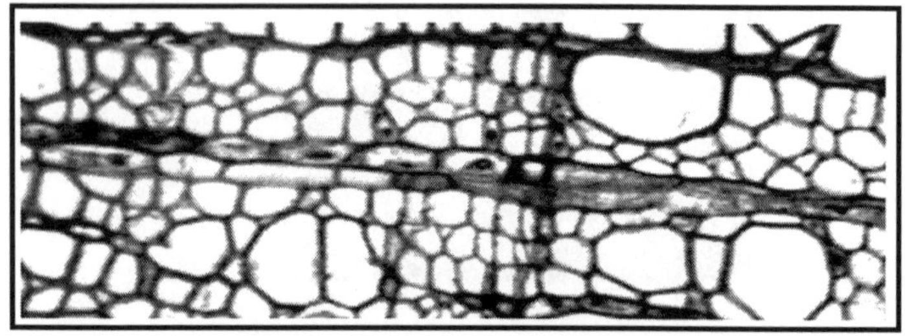

Fig. Closeup of a Cross-section of Basswood Growth Ring.

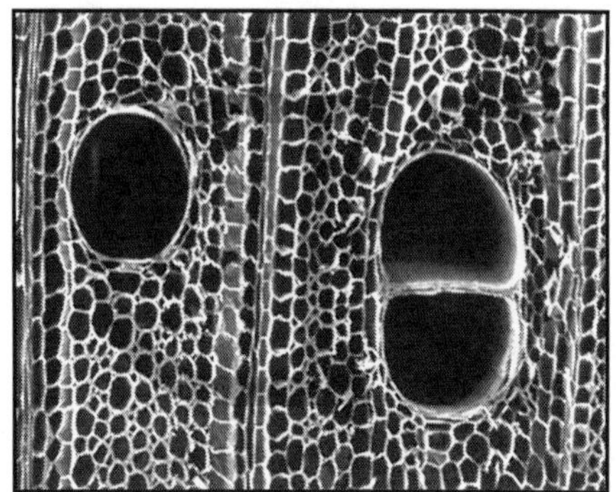

Fig. Balsa Wood (Cross Section) Showing Large Conductive Elements.

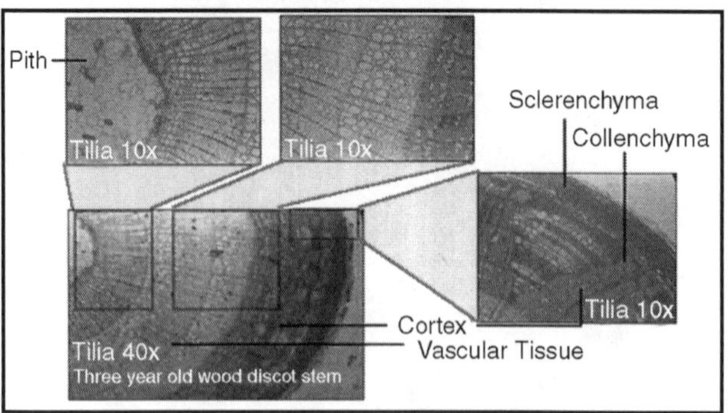

Fig. The Stem of Basswood.

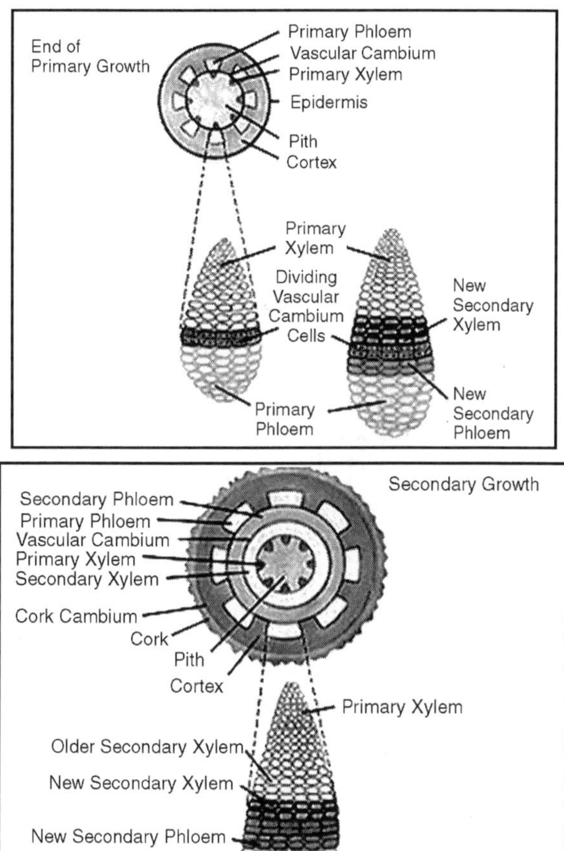

Fig. The Formation of Secondary Growth.

Monocots usually don't have secondary growth. Some, such as bamboo and palm trees, have secondary growth. Monocot secondary growth differs from dicot secondary growth in that new bundles are formed at the edge of the stem. These new bundles are close together, providing support for the stem.

The Leaf

The leaf consists of the (generally) flat blade, one or more leaf veins, a petiole, and usually an axillary bud. The petiole can be long (as in celery and bok-choy) or short (as in cabbage and lettuce). Leaves may be simple or compound: simple leaves have a single subdivision or leaflet, compound leaves have more than one leaflet. Leaves attach to stems at nodes (internodes are the spaces between nodes). Leaf phyllotaxy is the pattern exhibited (spiral, opposite, alternate, whorled) of leaf attachment to a stem.

PLASTICITY AND TOTIPOTENCY

Two concepts, plasticity and totipotency, are central to understanding plant cell culture and regeneration. Plants, due to their sessile nature and long life span, have developed a greater ability to endure extreme conditions and predation than have animals. Many of the processes involved in plant growth and development adapt to environmental conditions. This plasticity allows plants to alter their metabolism, growth and development to best suit their environment.

Particularly important aspects of this adaptation, as far as plant tissue culture and regeneration are concerned, are the abilities to initiate cell division from almost any tissue of the plant and to regenerate lost organs or undergo different developmental pathways in response to particular stimuli. When plant cells and tissues are cultured in vitro they generally exhibit a very high degree of plasticity, which allows one type of tissue or organ to be initiated from another type. In this way, whole plants can be subsequently regenerated.

This regeneration of whole organisms depends upon the concept that all plant cells can, given the correct stimuli, express the total genetic potential of the parent plant. This maintenance of genetic potential is called 'totipotency'. Plant cell culture and regeneration do, in fact, provide the most compelling evidence for totipotency. In practical terms though, identifying the culture conditions and stimuli required to manifest this totipotency can be extremely difficult and it is still a largely empirical process.

THE CULTURE ENVIRONMENT

When cultured in vitro, all the needs, both chemical and physical, of the plant cells have to met by the culture vessel, the growth medium and the external environment (light, temperature, etc.). The growth medium has to supply all the essential mineral ions required for growth and development. In many cases (as the biosynthetic capability of cells cultured in vitro may not replicate that of the parent plant), it must also supply additional organic supplements such as amino acids and vitamins.

Many plant cell cultures, as they are not photosynthetic, also require the addition of a fixed carbon source in the form of a sugar (most often sucrose). One other vital component that must also be supplied is water, the principal biological solvent. Physical factors, such as temperature, pH, the gaseous environment, light (quality and duration) and osmotic pressure, also have to be maintained within acceptable limits.

PLANT CELL CULTURE MEDIA

Culture media used for the in vitro cultivation of plant cells are composed of three basic components:

- Essential elements, or mineral ions, supplied as a complex mixture of salts;
- An organic supplement supplying vitamins and/or amino acids;
- A source of fixed carbon; usually supplied as the sugar sucrose.

Table: Some of the Elements Important for Plant Nutrition and their Physiological Function.

Element	*Function*
Nitrogen	Component of proteins, nucleic acids and some coenzymes Element required in greatest amount
Potassium	Regulates osmotic potential, principal inorganic cation
Calcium	Cell wall synthesis, membrane function, cell signalling
Magnesium	Enzyme cofactor, component of chlorophyll
Phosphorus	Component of nucleic acids, energy transfer, component of intermediates in respiration and photosynthesis
Sulphur	Component of some amino acids (methionine, cysteine) and some cofactors
Chlorine	Required for photosynthesis
Iron	Electron transfer as a component of cytochromes
Manganese	Enzyme cofactor
Cobalt	Component of some vitamins
Copper	Enzyme cofactor, electron-transfer reactions
Zinc	Enzyme cofactor, chlorophyll biosynthesis
Molybdenum	Enzyme cofactor, component of nitrate reductase

For practical purposes, the essential elements are further divided into the following categories:

- Macroelements (or macronutrients);
- Microelements (or micronutrients); and
- An iron source.

Media Components

It is useful to briefly consider some of the individual components of the stock solutions.

Macroelements

As is implied by the name, the stock solution supplies those elements required in large amounts for plant growth and development. Nitrogen, phosphorus, potassium, magnesium, calcium and sulphur (and carbon, which is added separately) are usually regarded as macroelements. These elements usually comprise at least 0.1 per cent of the dry weight of plants.

Nitrogen is most commonly supplied as a mixture of nitrate ions (from the KNO_3) and ammonium ions (from the NH_4NO_3). Theoretically, there is an advantage in supplying nitrogen in the form of ammonium ions, as nitrogen must be in the reduced form to be incorporated into macromolecules. Nitrate ions therefore need to be reduced before incorporation. However, at high concentrations, ammonium ions can be toxic to plant cell cultures and uptake of ammonium ions from the medium causes acidification of the medium. In order to use ammonium ions as the sole nitrogen source, the medium needs to be buffered.

High concentrations of ammonium ions can also cause culture problems by increasing the frequency of vitrification (the culture appears pale and 'glassy' and is usually unsuitable for further culture). Using a mixture of nitrate and ammonium ions has the advantage of weakly buffering the medium as the uptake of nitrate ions causes OH– ions to be excreted. Phosphorus is usually supplied as the phosphate ion of ammonium, sodium or potassium salts. High concentrations of phosphate can lead to the precipitation of medium elements as insoluble phosphates.

Microelements

These elements are required in trace amounts for plant growth and development, and have many and diverse roles. Manganese, iodine, copper, cobalt, boron, molybdenum, iron and zinc usually comprise the microelements, although other elements such as nickel and aluminium are frequently found in some formulations.

Iron is usually added as iron sulphate, although iron citrate can also be used. Ethylenediaminetetraacetic acid (EDTA) is usually used in conjunction with the iron sulphate. The EDTA complexes with the iron so as to allow the slow and continuous release of iron into the medium. Uncomplexed iron can precipitate out of the medium as ferric oxide.

Organic Supplements

Only two vitamins, thiamine (vitamin B1) and myoinositol (considered a B vitamin) are considered essential for the culture of plant cells in vitro. However, other vitamins are often added to plant cell culture media for historical reasons.

Amino acids are also commonly included in the organic supplement. The most frequently used is glycine (arginine, asparagine, aspartic acid, alanine, glutamic acid, glutamine and proline are also used), but in many cases its inclusion is not essential. Amino acids provide a source of reduced nitrogen and, like ammonium ions, uptake causes acidification of the medium. Casein hydrolysate can be used as a relatively cheap source of a mix of amino acids.

Carbon Source

Sucrose is cheap, easily available, readily assimilated and relatively stable and is therefore the most commonly used carbon source. Other carbohydrates (such as glucose, maltose, galactose and sorbitol) can also be used, and in specialised circumstances may prove superior to sucrose.

Gelling Agents

Media for plant cell culture in vitro can be used in either liquid or 'solid' forms, depending on the type of culture being grown. For any culture types that require the plant cells or tissues to be grown on the surface of the medium, it must be solidified (more correctly termed 'gelled'). Agar, produced from seaweed, is the most common type of gelling agent, and is ideal for routine applications.

However, because it is a natural product, the agar quality can vary from supplier to supplier and from batch to batch. For more demanding applications, a range of purer (and in some cases, considerably more expensive) gelling agents are available. Purified agar or agarose can be used, as can a variety of gellan gums.

These components, then, are the basic 'chemical' necessities for plant cell culture media. However, other additions are made in order to manipulate the pattern of growth and development of the plant cell culture.

PLANT GROWTH REGULATORS

We have already briefly considered the concepts of plasticity and totipotency. The essential point as far as plant cell culture is concerned is that, due to this plasticity and totipotency, specific media manipulations can be used to direct the development of plant cells in culture.

Plant growth regulators are the critical media components in determining the developmental pathway of the plant cells. The plant growth regulators used most commonly are plant hormones or their synthetic analogues.

Classes of Plant Growth Regulators

There are five main classes of plant growth regulator used in plant cell culture, namely:

- Auxins;
- Cytokinins;
- Gibberellins;
- Abscisic acid;
- Ethylene.

Each class of plant growth regulator will be briefly looked at.

Auxins

Auxins promote both cell division and cell growth The most important naturally occurring auxin is IAA (indole-3-acetic acid), but its use in plant cell culture media is limited because it is unstable to both heat and light.

Occasionally, amino acid conjugates of IAA (such as indole-acetyl-L-alanine and indole-acetyl-L-glycine), which are more stable, are used to partially alleviate the problems associated with the use of IAA. It is more common, though, to use stable chemical analogues of IAA as a source of auxin in plant cell culture media. 2,4-Dichlorophenoxyacetic acid (2,4-D) is the most commonly used auxin and is extremely effective in most circumstances. Other auxins are available, and some may be more effective or 'potent' than 2,4- D in some instances.

Table: Commonly used Auxins, their Abbreviation and Chemical Name

Abbreviation/name	*Chemical name*
2,4 – D	2,4 – Dichlorophenoxyacetic acid
2,4,5 – T	2,4,5 – Trichlorophenoxyacetic acid
Dicamba	2-Methoxy – 3,6 – Dichlorobenzoic acid
IAA	Indole – 3 – Acetic Acid
IBA	Indole – 3 – Butyric Acid
MCPA	2-Methyl – 4 – Chlorophenoxyacetic Acid
NAA	1 – Naphthylacetic Acid
NOA	2 – Naphthyloxyacetic Acid
Picloram	4 – Amino – 2,5,6 – Trichloropicolinic acid

Table: Commonly used Cytokinins, their Abbreviation and Chemical Name

Abbreviation/name	*Chemical name*
BAPa	6-benzylaminopurine
2iP (IPA)b	[N6-(2-isopentyl)adenine]
Kinetina	6-furfurylaminopurine
Thidiazuronc	1-phenyl-3-(1,2,3-thiadiazol-5-yl) urea
Zeatinb	4-hydroxy-3-methyl-trans-2-butenylaminopurine

Cytokinins

Cytokinins promote cell division. Naturally occurring cytokinins are a large group of structurally related (they are purine derivatives) compounds. Of the naturally occurring cytokinins, two have some use in plant tissue culture media.

These are zeatin and 2iP (2-isopentyl adenine). Their use is not widespread as they are expensive (particularly zeatin) and relatively unstable.

The synthetic analogues, kinetin and BAP (benzylaminopurine), are therefore used more frequently. Non-purine-based chemicals, such as substituted phenylureas, are also used as cytokinins in plant cell culture media. These substituted phenylureas can also substitute for auxin in some culture systems.

Gibberellins

There are numerous, naturally occurring, structurally related compounds termed 'gibberellins'. They are involved in regulating cell elongation, and are agronomically important in determining plant height and fruit-set. Only a few of the gibberellins are used in plant tissue culture media, GA3 being the most common.

Abscisic Acid

Abscisic acid (ABA) inhibits cell division. It is most commonly used in plant tissue culture to promote distinct developmental pathways such as somatic embryogenesis.

Ethylene

Ethylene is a gaseous, naturally occurring, plant growth regulator most commonly associated with controlling fruit ripening in climacteric fruits, and its use in plant tissue culture is not widespread. It does, though, present a particular problem for plant tissue culture. Some plant cell cultures produce ethylene, which, if it builds up sufficiently, can inhibit the growth and development of the culture. The type of culture vessel used and its means of closure affect the gaseous exchange between the culture vessel and the outside atmosphere and thus the levels of ethylene present in the culture.

PLANT GROWTH REGULATORS AND TISSUE CULTURE

Generalisations about plant growth regulators and their use in plant cell culture media have been developed from initial observations made in the 1950s. There is, however, some considerable difficulty in predicting the effects of plant growth regulators: this is because of the great differences in culture response between species, cultivars and even plants of the same cultivar grown under different conditions.

However, some principles do hold true and have become the paradigm on which most plant tissue culture regimes are based. Auxins and cytokinins are the most widely used plant growth regulators in plant tissue culture and are usually used together, the ratio of the auxin to the cytokinin determining the type of culture established or regenerated. A high auxin to cytokinin ratio generally favours root formation, whereas a high cytokinin to auxin ratio favours shoot formation. An intermediate ratio favours callus production.

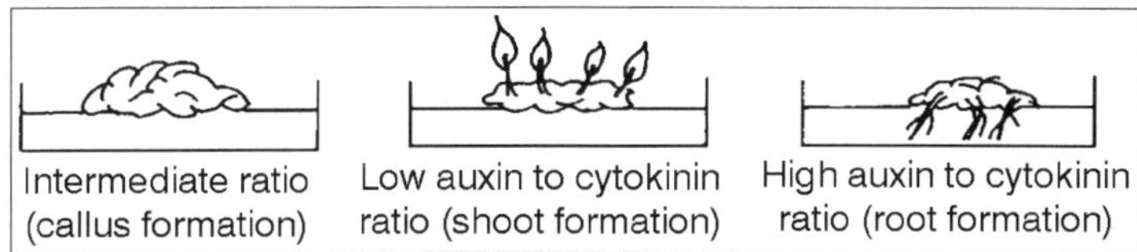

Fig. The Effect of Different Ratios of Auxin to Cytokinin on the Growth and Morphogenesis of Callus. High Auxin to Cytokinin Ratios Promote Root Development, Low Ratios Promote Shoot Development. Intermediate Ratios Promote Continued Growth of the Callus without Differentiation

CULTURE TYPES

Cultures are generally initiated from sterile pieces of a whole plant. These pieces are termed 'explants', and may consist of pieces of organs, such as leaves or roots, or may be specific cell types, such as pollen or endosperm. Many features of the explant are known to affect the efficiency of culture initiation. Generally, younger, more rapidly growing tissue (or tissue at an early stage of development) is most effective. Several different culture types most commonly used in plant transformation studies will now be examined in more detail.

CALLUS

Explants, when cultured on the appropriate medium, usually with both an auxin and a cytokinin, can give rise to an unorganised, growing and dividing mass of cells. It is thought that any plant tissue can be used as an explant, if the correct conditions are found. In culture, this proliferation can be maintained more or less indefinitely, provided that the callus is subcultured on to fresh medium periodically. During callus formation there is some degree of dedifferentiation (i.e. the changes that occur during development and specialisation are, to some extent, reversed), both in morphology (callus is usually composed of unspecialised parenchyma cells) and metabolism.

One major consequence of this dedifferentiation is that most plant cultures lose the ability to photosynthesise. This has important consequences for the culture of callus tissue, as the metabolic profile will probably not match that of the donor plant. This necessitates the addition of other components—such as vitamins and, most importantly, a carbon source—to the culture medium, in addition to the usual mineral nutrients.

Callus culture is often performed in the dark (the lack of photosynthetic capability being no drawback) as light can encourage differentiation of the callus. During long-term culture, the culture may lose the requirement for auxin and/or cytokinin. This process, known as 'habituation', is common in callus cultures from some plant species (such as sugar beet). Callus cultures are extremely important in plant biotechnology. Manipulation of the auxin to cytokinin ratio in the medium can lead to the development of shoots, roots or somatic embryos from which whole plants can subsequently be produced. Callus cultures can also be used to initiate cell suspensions, which are used in a variety of ways in plant transformation studies.

CELL-SUSPENSION CULTURES

Callus cultures, broadly speaking, fall into one of two categories: compact or friable. In compact callus the cells are densely aggregated, whereas in friable callus the cells are only loosely associated with each other and the callus becomes soft and breaks apart easily. Friable callus provides the inoculum to form cell-suspension cultures. Explants from some plant species or particular cell types tend not to form friable callus, making cell-suspension initiation a difficult task. The friability of callus can sometimes be improved by manipulating the medium components or by repeated subculturing. The friability of the callus can also sometimes be improved by culturing it on 'semi-solid' medium (medium with a low concentration of gelling agent).

When friable callus is placed into a liquid medium (usually the same composition as the solid medium used for the callus culture) and then agitated, single cells and/or small clumps of cells are released into the medium. Under the correct conditions, these released cells continue to grow and divide, eventually producing a cell-suspension culture. A relatively large inoculum should be used when initiating cell suspensions so that the released cell numbers build up quickly. The inoculum should not be too large though, as toxic products released from damaged or stressed cells can build up to lethal levels. Large cell clumps can be removed during subculture of the cell suspension. Cell suspensions can be maintained relatively simply as batch cultures in conical flasks. They are continually cultured by repeated subculturing into fresh medium. This results in dilution of the suspension and the initiation of another batch growth cycle. The degree of dilution during subculture should be determined empirically for each culture. Too great a degree of dilution will result in a greatly extended lag period or, in extreme cases, death of the transferred cells.

After subculture, the cells divide and the biomass of the culture increases in a characteristic fashion, until nutrients in the medium are exhausted and/or toxic by-products build up to inhibitory levels—this is called the 'stationary phase'. If cells are left in the stationary phase for too long, they will die and the culture will be lost. Therefore, cells should be transferred as they enter the stationary phase. It is therefore important that the batch growth-cycle parameters are determined for each cell-suspension culture.

PROTOPLASTS

Protoplasts are plant cells with the cell wall removed. Protoplasts are most commonly isolated from either leaf mesophyll cells or cell suspensions, although other sources can be used to advantage. Two general approaches to removing the cell wall (a difficult task without damaging the protoplast) can be taken—mechanical or enzymatic isolation.

Mechanical isolation, although possible, often results in low yields, poor quality and poor performance in culture due to substances released from damaged cells. Enzymatic isolation is usually carried out in a simple salt solution with a high osmoticum, plus the cell wall degrading enzymes. It is usual to use a mix of both cellulase and pectinase enzymes, which must be of high quality and purity.

Protoplasts are fragile and easily damaged, and therefore must be cultured carefully. Liquid medium is not agitated and a high osmotic potential is maintained, at least in the initial stages. The liquid medium must be shallow enough to allow aeration in the absence of agitation. Protoplasts can be plated out on to solid medium and callus produced. Whole plants can be regenerated by organogenesis or somatic embryogenesis from this callus. Protoplasts are ideal targets for transformation by a variety of means.

ROOT CULTURES

Root cultures can be established in vitro from explants of the root tip of either primary or lateral roots and can be cultured on fairly simple media. The growth of roots in vitro is potentially unlimited, as roots are indeterminate organs. Although the establishment of root cultures was one of the first achievements of modern plant tissue culture, they are not widely used in plant transformation studies.

SHOOT TIP AND MERISTEM CULTURE

The tips of shoots (which contain the shoot apical meristem) can be cultured in vitro, producing clumps of shoots from either axillary or adventitious buds. This method can be used for clonal propagation. Shoot meristem cultures are potential alternatives to the more commonly used methods for cereal regeneration as they are less genotype-dependent and more efficient (seedlings can be used as donor material).

EMBRYO CULTURE

Embryos can be used as explants to generate callus cultures or somatic embryos. Both immature and mature embryos can be used as explants. Immature, embryo-derived embryogenic callus is the most popular method of monocot plant regeneration.

MICROSPORE CULTURE

Haploid tissue can be cultured in vitro by using pollen or anthers as an explant. Pollen contains the male gametophyte, which is termed the 'microspore'. Both callus and embryos can be produced from pollen. Two main approaches can be taken to produce in vitro cultures from haploid tissue. The first method depends on using the anther as the explant. Anthers (somatic tissue that surrounds and contains the pollen) can be cultured on solid medium (agar should not be used to solidify the medium as it contains inhibitory substances). Pollen-derived embryos are subsequently produced via dehiscence of the mature anthers.

The dehiscence of the anther depends both on its isolation at the correct stage and on the correct culture conditions. In some species, the reliance on natural dehiscence can be circumvented by cutting the wall of the anther, although this does, of course, take a considerable amount of time. Anthers can also be cultured in liquid medium, and pollen released from the anthers can be induced to form embryos, although the efficiency of plant regeneration is often very low. Immature pollen can also be extracted from developing anthers and cultured directly, although this is a very time-consuming process.

Both methods have advantages and disadvantages. Some beneficial effects to the culture are observed when anthers are used as the explant material. There is, however, the danger that some of the embryos produced from anther culture will originate from the somatic anther tissue rather than the haploid microspore cells. If isolated pollen is used there is no danger of mixed embryo formation, but the efficiency is low and the process is time-consuming. In microspore culture, the condition of the donor plant is of critical importance, as is the timing of isolation. Pretreatments, such as a cold treatment, are often found to increase the efficiency.

These pretreatments can be applied before culture, or, in some species, after placing the anthers in culture. Plant species can be divided into two groups, depending on whether they require the addition of plant growth regulators to the medium for pollen/anther culture; those that do also often require organic supplements, e.g. amino acids. Many of the cereals (rice, wheat, barley and maize) require medium supplemented with plant growth regulators for pollen/anther culture. Regeneration from microspore explants can be obtained by direct embryogenesis, or via a callus stage and subsequent embryogenesis.

Haploid tissue cultures can also be initiated from the female gametophyte (the ovule). In some cases, this is a more efficient method than using pollen or anthers. The ploidy of the plants obtained from haploid cultures may not be haploid. This can be a consequence of chromosome doubling during the culture period. Chromosome doubling (which often has to be induced by treatment with chemicals such as colchicine) may be an advantage, as in many cases haploid plants are not the desired outcome of regeneration from haploid tissues. Such plants are often referred to as 'di-haploids', because they contain two copies of the same haploid genome.

PLANT REGENERATION

Having looked at the main types of plant culture that can be established in vitro, we can now look at how whole plants can be regenerated from these cultures. In broad terms, two methods of plant regeneration are widely used in plant transformation studies, i.e. somatic embryogenesis and organogenesis.

SOMATIC EMBRYOGENESIS

In somatic (asexual) embryogenesis, embryo-like structures, which can develop into whole plants in a way analogous to zygotic embryos, are formed from somatic tissues. These somatic embryos can be produced either directly or indirectly. In direct somatic embryogenesis, the embryo is formed directly from a cell or small group of cells without the production of an intervening callus. Though common from some tissues (usually reproductive tissues such as the nucellus, styles or pollen), direct somatic embryogenesis is generally rare in comparison with indirect somatic embryogenesis.

In indirect somatic embryogenesis, callus is first produced from the explant. Embryos can then be produced from the callus tissue or from a cell suspension produced from that callus. Somatic embryogenesis usually proceeds in two distinct stages. In the initial stage (embryo initiation), a high concentration of 2,4-D is used. In the second stage (embryo production) embryos are produced in a medium with no or very low levels of 2,4-D

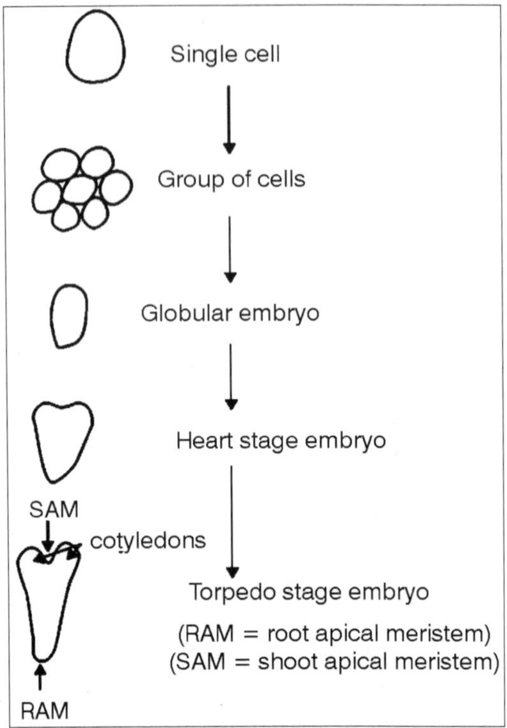

Fig. A Schematic Representation of the Sequential Stages of Somatic Embryo Development

Somatic embryos may develop from single cells or from a small group of cells. Repeated cell divisions lead to the production of a group of cells that develop into an organised structure known as a 'globular-stage embryo'. Further development results in heart- and torpedostage embryos, from which plants can be regenerated. Zygotic embryos undergo a fundamentally similar development through the globular (which is formed after the 16- cell stage), heart and torpedo stages. Polarity is established early in embryo development. Signs of tissue differentiation become apparent at the globular stage and apical meristems are apparent in heart-stage embryos.

SOMATIC EMBRYOGENESIS

INDIRECT SOMATIC EMBRYOGENESIS IN CARROT (DAUCUS CAROTA)

A callus can be established from explants from a wide range of carrot tissues by placing the explant on solid medium (e.g. Murashige and Skoog (MS)) containing 2,4-D (1mgl–1). This callus can be used to produce a cell suspension by placing it in agitated liquid MS medium containing 2,4-D (1mgl–1). This cell suspension can be maintained by repeated subculturing into 2,4-D-containing medium. Removal of the old 2,4-D-containing medium and replacement with fresh medium containing abscisic acid (0.025mgl–1) results in the production of embryos.

Direct Somatic Embryogenesis from Alfalfa (Medicago Falcata)

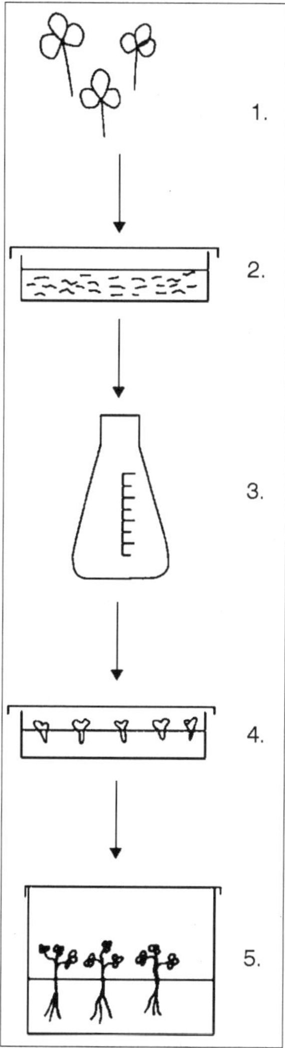

Fig. Direct somatic embryogenesis in alfalfa

Young trifoliate leaves are used as the explant. These are removed from the plant and chopped into small pieces. The pieces are washed in a plant growth regulator-free medium and placed in liquid medium (B5) supplemented with 2,4-D (4mgl–1), kinetin (0.2mgl–1), adenine (1mgl–1) and glutathione

- Explants are removed from plants grown in vitro.
- Explants are placed in liquid medium for embryo induction.
- Embryos develop to the globular stage in liquid medium supplemented with maltose and polye-thylene glycol.
- Embryos mature on gelled medium containing abscisic acid (ABA).
- Embryos develop into plants on solid medium.

The cultures are maintained in agitated liquid medium for about 10–15 days. Washing the explants and replacing the old medium with B5 medium supplemented with maltose and polyethylene glycol results in the development of the somatic embryos. These somatic embryos can be matured on solid medium containing abscisic acid.

Note that in both cases, although the production of somatic embryos from alfalfa necessitates the use of more complicated media, the production of embryos is fundamentally a two-step process. The initial medium, which contains 2,4-D, is replaced with a medium that does not contain 2,4-D. In many systems it has been found that somatic embryogenesis is improved by supplying a source of reduced nitrogen, such as specific amino acids or casein hydrolysate.

CEREAL REGENERATION

The principal method adopted for the tissue culture and regeneration of a wide range of cereal species is somatic embryogenesis, using cultures initiated from immature zygotic embryos. Embryogenic callus is normally initiated by placing the immature embryo on to a medium containing 2,4-D. Shoot regeneration is initiated by placing the embryogenic callus on a medium with BAP (with or without 2,4-D). These shoots can be subsequently rooted. The medium used for induction of embryogenic callus is usually a modified MS (for Triticeae) or N6 (for rice and maize). Maltose is often used as the carbon source in preference to sucrose, and additional organic supplements (such as specific amino acids, yeast extract and/or casein hydrolysate) are common.

The isolation and culture of immature embryos is, however, a labour-intensive and relatively expensive procedure. An additional problem is the small target size if the immature embryos are to used for biolistic transformation. Alternatives are therefore being sought. One alternative is to use mature embryos (or seeds) as the explant to initiate embryogenic callus. This approach has been successfully applied to several cereal species such as rice and oats. The culture techniques and media used for culture establishment and regeneration from mature embryo/seed-derived cultures are fundamentally the same as those used for cultures initiated from immature embryos.

THE IMPORTANCE OF GENOTYPE

The major influence on tissue-culture response appears to be genetic, with culture requirements varying between species and cultivars. Model genotypes that responded well to culture in vitro were initially used in plant transformation studies. However, most of the model genotypes used were not elite, commercial cultivars.

The commercial cultivars tended to respond poorly to culture in vitro. One of the main aims is therefore to identify the components that make up a widely applicable, optimal culture regime. Many factors have been investigated for their ability to improve the culture response from elite cultivars, including media components (such as alternative carbon sources, macro- and microelement concentrations and composition), media preparation method and donor plant condition and growth conditions.

ORGANOGENESIS

Somatic embryogenesis relies on plant regeneration through a process analogous to zygotic embryo germination. Organogenesis relies on the production of organs, either directly from an explant or from a callus culture. There are three methods of plant regeneration via organogenesis.

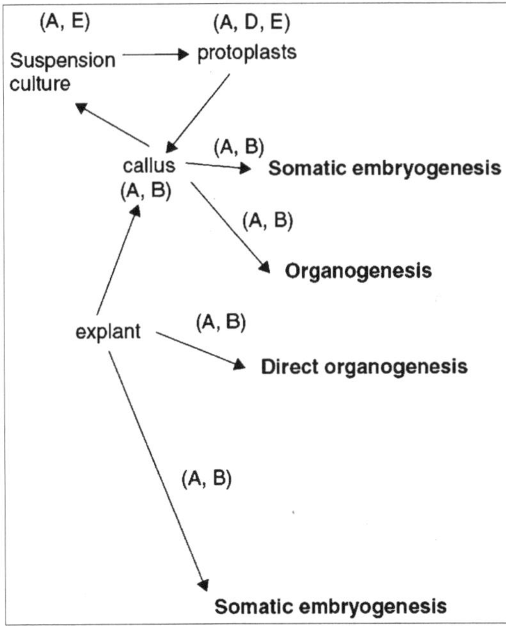

Fig. A Simplified Scheme for the Integration of Plant Tissue Culture into Plant Transformation Protocols

The first two methods depend on adventitious organs arising either from a callus culture or directly from an explant. Alternatively, axillary bud formation and growth can also be used to regenerate whole plants from some types of tissue culture. Organogenesis relies on the inherent plasticity of plant tissues, and is regulated by altering the components of the medium. In particular, it is the auxin to cytokinin ratio of the medium that determines which developmental pathway the regenerating tissue will take. It is usual to induce shoot formation by increasing the cytokinin to auxin ratio of the culture medium. These shoots can then be rooted relatively simply.

An explant can be a variety of tissues, depending on the particular plant species being cultured. The explant can be used to initiate a variety of culture types, depending on the explant used. Regeneration by either organogenesis or somatic embryogenesis results in the production of whole plants. Different culture types and regeneration methods are amenable to different transformation protocols.

The transformation protocols highlighted in this figure are:

- Agrobacterium-mediated;
- Biolistic transformation;
- Direct DNA uptake
- Electroporation.

Different combinations of culture type and transformation protocol are used depending on the plant species and cultivar being used. In some species a variety of culture types and regeneration methods can be used, which enables a wide variety of transformation protocols to be utilised. In other species there is effectively no choice over culture type and/or regeneration method, which can limit the transformation protocols that are applicable.

TISSUE CULTURE TECHNIQUES

The medicinal value of the extract from P africana bark for the treatment of benign prostatic hyperplasia, a common condition in elderly men has led to international trade worth approximately USD 220M per year, in the final pharmaceutical product. Approximately 4000t of bark is collected annually through destructive harvesting of trees from natural stands to meet this demand. This bark exploitation has caused serious damage to wild populations of Prunus leading to concerns on the long term sustainability of harvesting and conservation of this tree species.

This research aims at conserving the Prunus (ex-situ) through provenance collections of overexploited Prunus germplasm, particularly where commercial bark harvesting is taking place and their subsequent conservation in conservation stands. It also aims at the rapid multiplication of selected Prunus africana germplasm through tissue culture techniques for on farm planting by farmers. This will provide an alternative to the destructive sampling of wild stocks as well as help arrest the rapid erosion of overexploited Prunus germplasm.

Multiplication through tissue culture techniques will be done for selected trees which are, high active-ingredient yielding, fast growing, and of good form, using indigenous knowledge. The expected outputs from the research are conservation of selected overexploited Prunus germplasm from various provenances, through the establishment of gene banks, distribution of already characterized and evaluated, tissue culture multiplied germplasm for planting by farmers and documentation of the results and information for dissemination to stakeholders. To date successful plantations and enrichment plantings have been limited to Kenya where over a period of 90 years, Prunus africana has been planted by the Forest department for timber. However, their planting material is often of unknown origin and of narrow genetic base.

They are therefore a limited utility for conservation purposes. A study was carried by Cunningham and Mbenkum, to investigate the economic feasibility of different planting systems enrichment planting, small scale farming and plantations for Prunus as an alternative to the current overexploitation of wild populations in Cameroon. Some of the recommendations made were that low technology propagators at field sites would boost the number of young seedlings being produced, and research needs to be conducted for selection of fast growing, high active yielding Prunus africana cultivars.

If more attention is placed on the origin and genetic diversity of material, enrichment plantings and plantations can serve a useful conservation function and act as a source of planting material for other users. This research therefore aims to select in natural forests and farmer's fields overexploited Prunus germplasm, and plant them in gene banks, for their conservation. Due to the intermediate nature of the seed, propagating the tree through seeds may be difficult. Furthermore, it takes 15 years for the trees to seed. Propagating Prunus through tissue culture is a rapid method of multiplying the Prunus africana and is a faster way of achieving our objectives of conservation.

Prunus Africana is popularly known as Pygeum, and is a large evergreen tree that grows in the afromontane of Africa. It is the only species of Genus Prunus native to Africa and can grow to a height of up to 40 meters. It has pendulous branches with thick oblong- shaped, leather like, mat colored leaves and creamy white flowers. The fruit (drupe) resembles a cherry when ripe. Traded internationally and harvested from the wild, the Prunus is hardly a minor forest product.

The tree is valued for its medicinal extract from its bark for the treatment of benign prostatic hyperplasia, a common condition in elderly men. Annually, approximately 4000t of bark are harvested primarily from Cameroon, Equatorial Guinea, Kenya and Madagascar to meet this demand.This bark exploitation has caused serious damage to wild populations of Prunus including trees inside forests of high conservation value, leading to concerns on the long term sustainability of harvesting and conservation of this tree species.

Diversity of other uses of this tree include making of handles for axes and hoes, firewood and poles. A serious constraint in the expansion of on farm tree planting is seed availability. Here a key difficulty is the intermediate nature of the seed. Furthermore seed yields fluctuate widely between years. Seed shortage is likely to be exacerbated in future as the size of natural populations diminishes. The approximate time to first flowering and fruiting is 15 years. Raising Prunus through tissue culture techniques will therefore accelerate their production. The transfer of germplasm from the wild into on farm niches helps to preserve valuable genetic resources particularly if attention is paid to the origin and diversity of cultivated material. In addition, cultivation of endangered forests species takes pressure off their natural resource base, thus promoting the conservation of natural forests.

Objectives of the Project

- To gather through explorations and collection expeditions in different ecological zones, overexploited Prunus germplasm as well as meaningful and rich stock of Prunus with superior characteristics.
- Rapid multiplication (through Tissue culture techniques) of both the heavily exploited germplasm and the Prunus with superior characteristics, for establishment of conservation stands and their distribution to farmers.

Hypothesis

Conservation of Prunus africana can be enhanced through tissue culture techniques.

Expected out Puts

- Conservation of the overexploited Prunus germplasm of different provenances, in established conservation stands
- Continued availability of the selected Prunus germplasm through their continuous multiplication.
- Documented research results for ease of dissemination and retrieval.

Selection

Three sites have been chosen to be areas in which selection of Prunus africana will be done. These are Mt. Kenya forest, Mt. Elgon forest and the Abardares. These sites were chosen as they are areas in which the tree is known to occur naturally and is heavily overexploited. Random selection of trees will be done in the chosen respective provenances with the help of forest officers and farmers in the communities with indigenous knowledge.Selection will be done for trees that are high ingredient yielding, fast growing, early seeding, and have good form i.e. trees that are straight, with no mechanical damage no forking or kinking, no dwafing or diseased.

Germplasm will be collected from trees which have passed selection criteria as well as from trees that have been overexploited and are becoming extinct. Measurements to be taken for the selected trees are diameter at breat height and height of the trees. Other data to be noted on site include, approximate age and location of the tree. Branches or coppices will be collected from 25 randomly chosen trees per provenance using secateurs. They will be packed according to provenances and carefully labeled using marker pens. The packaging will consist of polythene bags containing water, which will be stored in cooler boxes. Seeds will be collected from 25 randomly chosen and supposedly unrelated individuals using a seed harvester, packed in khaki bags which will also be labeled using a marker pen.

The collected coppices will be brought to the lab for propagation through tissue culture techniques. Bark samples will also be collected from the selected trees and packed in Khaki bags, which will also be labeled. The bark samples will also be taken to selected laboratory for chemical analysis. Collected seeds will be germinated in seed beds and then transplanted into potted tubes containing soil media. They will be planted in conservation stands once they are ready for planting.

Field Trial Design (Conservation Stands)

The sites for planting the propagated material for both seed and tissue culture from the tree provenances will be identified. The land will be cleared, ploughed and planting holes dug. The sites will be at isolation belts of between 300 to 500 meters from other natural vegetation that have Prunus trees.The propagated material ready for field planting, will be planted out in the field during the onset of the rains, using local labour. The material from the three provenances will be planted in each of the provenance. Each provenance will have a total of 533 trees and will be replicated 3 times across the trial design. Each replicate will therefore have 133 trees from each provenance. There will be 4 replicates for each of the provenance. Replicates will be used for planting prunus propagated from seed.

The site will be visited twice a year for monitoring and maintenance i.e. weeding. Data will be collected once a year. This includes data for survival, height, root collar, diameter and diameter at breast height for 10 randomly selected trees per plot of 133 trees. The data will be analysed at the nursery. Thinning will also be done when necessary.

Tissue Culture Procedures

The explants from the field will be stored in a refrigerator, while still in the polybags with water to keep them fresh. Tissue culture will be done progressively, one provenance at a time The explant to be used here are shoot tips form the Prunus.

However, a protocol will have to be worked out through trial and error, to establish which media combination will be successful in the propagation of the prunus. The culture media should have a pH of 5.7 +/- 0.1, before agar is added. This pH is achieved by measuring the media pH using the pH meter and adding HCL or NaOH to adjust the pH accordingly. Agar is always added last, and it should be mixed well, using heat, preferably in a microwave The prepared media will be discharged into test tubes (5mls per tube) using a syringe. The test tubes will be covered and packed in an autoclavable tray and placed in the autoclave. The autoclave will be set to 15 minutes at slow exhaust. When the autoclave cycle is complete and the pressure has reached zero, and temperature below 100°C, the autoclave door will be opened and the tray with tubes removed.

Surface Disinfection of Explant Material

The explants will be trimmed to size convenient to handle and placed in a container with little quantity of water to maintain the moisture until processing. The water in the container is then discarded. The material is then put into 70 per cent ethanol for 25 minutes, followed by 5-10 per cent sodium hypochlorite solution with few drops of tween for 20 minutes. The material is then rinsed with 3 changes of sterile distilled water. The explants are then kept in the last change of water. If the material is heavily contaminated, a double disinfection process with sodium hypochlorite may be used. The explants are now ready to take to the transfer chamber. This last step will be performed inside a laminar flow hood.

Starting the Explants (this is done in the Transfer/ Innoculation Room)

The transfer chamber should be ready with the walls and workspace sprayed and wiped with 70 per cent ethanol. Our hands should also have been washed with soap and water till the elbows, first, then sprayed with 70 per cent ethanol. There should be a container of 70 per cent ethanol to sterilize and rinse the instruments, I.e the blade (no 10) and forceps. Immerse the forceps and scapel for 30 seconds or more in the 70 per cent ehanol then burn off the etanol over the flame from the bunsen burner. With the forceps place an explant tip on the petri dish.

Place the forceps in the other hand to hold the tip while the first hand uses the scalpel to cut off 1 cm of the basipetal end of the shoot tip. Place the scapel in the70 per cent ethanol, move the forceps to the other hand. Grab a sterile test tube of medium with the other hand, hold it by the base. With the little finger of the hand holding the forceps, remove the cap and cradle it there while we use the forceps to firmlty lay the bud on the medium. Recap the test tube and seal with a piece of scotch tape (or Para film).

Growing

Place the test tubes in the planter tray (or other appropriate holder) and place the tray on a shelf under fluorescent light which is 8 - 10 inches above the top of the tubes, in the incubation room. Room temperature is o.k. Continous light is acceptable, but 16 hours light and 8 hours darkness is standard. Check daily for cotaminants. If any are found, sterilize tube and contents before discarding the contents in an autoclave Transfer the explant every two weeks or so until it is actively growing. In one to two months we should be able to divide the culture into two pieces, each of which is about 0.5 cm in diameter. Continue to divide and transfer until we have enough plantlets. The platlet should be singulated as we transfer to pre-rooting medium which has no hormone (or only IAA).

Transplanting

With the plantlets begin to root, perhaps two to four weeks, transplant them to a light artificial soil mix, such as peat/pearlite, in a seedling tray. Cover with clear plastic and place on a lighted shelf or in a shaded greenhouse. After two or three weeks, begin leaving the plastic off for a period of time each day. The time the plantlets are left uncovered should get longer each day, until after about a week, the cover can be left off completely. (Tissue cultured plantlets are more delicate than seedlings as the stomates remain open until they slowly adjust to normal humidity and light).

INTEGRATION OF PLANT TISSUE CULTURE

Various methods of plant regeneration are available to the plant biotechnologist. Some plant species may be amenable to regeneration by a variety of methods, but some may only be regenerated by one method. Not all plant tissue is suited to every plant transformation method, and not all plant species can be regenerated by every method. There is therefore a need to find both a suitable plant tissue culture/regeneration regime and a compatible plant transformation methodology.

Tissue culture and plant regeneration are an integral part of most plant transformation strategies, and can often prove to be the most challenging aspect of a plant transformation protocol. Key to success in integrating plant tissue culture into plant transformation strategies is the realisation that a quick (to avoid too many deleterious effects from somaclonal variation) and efficient regeneration system must be developed. However, this system must also allow high transformation efficiencies from whichever transformation technique is adopted. Not all regeneration protocols are compatible with all transformation techniques.

Some crops may be amenable to a variety of regeneration and transformation strategies, others may currently only be amenable to one particular protocol. Advances are being made all the time, so it is impossible to say that a particular crop will never be regenerated by a particular protocol. However, some protocols, at least at the moment, are clearly more efficient than others. Regeneration from immature embryo-derived somatic embryos is, for example, the favoured method for regenerating monocot species. General rules are sometimes difficult to make because of the variability of response to particular protocols of different plant species or even cultivars.

4

Micropropagation in Tissue Culture Technology

NODE MICROPROPAGATION

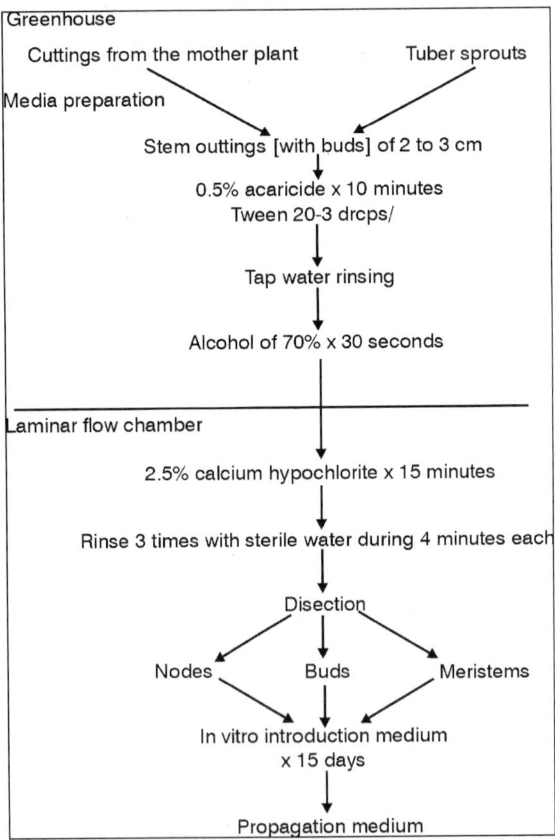

Fig. Procedure for in vitro Introducction

In vitro plantlets, which are free of pathogens, are used as initial material for potato and sweetpotato seed programs. The methods used in these micropropagation programs mainly depend on their production volume and the available infrastructure. In the case of potato micropropagation, the basic methods have been already described. They have been verified in many institutions and they are based on the rapid growth of individual node cuttings, or stems with multiple node cuttings. Afterwards, the basic micropropagation methods are described.

NODE MICROPROPAGATION

This method is based on the principle that the node of an in vitro plantlet placed in an appropriate culture medium will induce the development of the axillary bud, resulting in a new in vitro plantlet. This type of propagation promotes the development of a pre-existent morphological structure. The nutritional and hormonal condition of the medium breaks the dormancy of the axillary bud and promotes its rapid development. Callus formation and plant regeneration must be avoided because they tend to affect the genetic stability of the genotype. Under room-controlled conditions micropropagation is fast. Each node planted in a propagation medium will produce a plantlet which will occupy the full length of the test tube, after approximately four weeks for potato, and six weeks for sweetpotato. The resultant in vitro plantlets may be transplanted to in vitro conditions in small pots in the greenhouse.

MICROPROPAGATION BY NODE CUTTINGS IN A LIQUID MEDIUM

This technique is applied both with potato and sweetpotato to produce a large number of nodes rapidly. Stem cuttings with 5 to 8 nodes are prepared by removing both the apex and the root of the in vitro plant to be propagated. The stems are placed in the corresponding propagation liquid medium. It is also possible to use isolated nodes: the nodes will sprout and new plantlets will develop over a period of 3 to 4 weeks.

Micropropagation Procedure

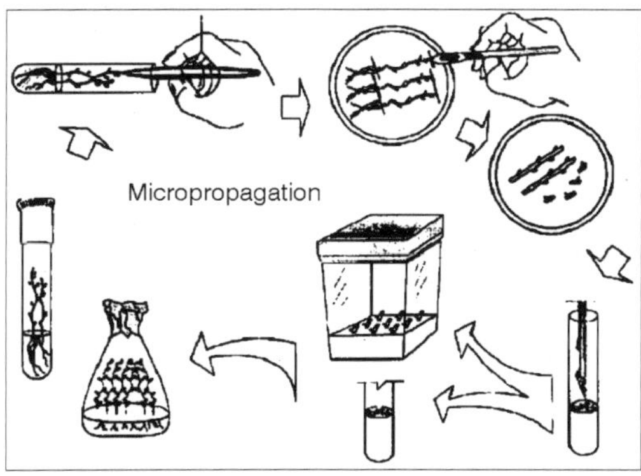

Fig. Potato Micropropagation Process Scheme

- Sterilize petri dishes (placed in paper bags or comets) and prepare the laminar flow chamber by disinfecting the internal surfaces with alcohol.
  Sterilize the tools with an instrument sterilizer and place them on a sterile dish.
- Open the tube, take off the plantlet and place it on a petri dish with the help of forceps.
- Remove the leaves and cut the nodes.
- Open a tube containing fresh sterile medium and place a node inside, trying to plunge it slightly into the medium with the bud up. Close the tube.
- Seal the tube with a gas-permeable plastic tape (parafilm or saram wrap) and label it correctly. It is recommended to place two explants in 16 x 125 mm tubes, three in 18 x 150 mm tubes, five in 25 x 150 mm tubes, and 20-30 in magenta vessels.

MEDIUM TERM GERMPLASM MAINTENANCE

The in vitro germplasm maintenance under normal growth conditions requires a series of transfers of the plantlets into a fresh medium. This leads to a consumption of time, increases the possibility of losses because of material contamination during successive sub-cultures, causes a loss of material by human error or failure of some equipment, and demands more labour A way to avoid these problems is through limitation, restriction or inhibition of growth. This approach consists of growth speed reduction by modifying the physical or chemical conditions of the culture, and it is effective for a short or medium term period.

IN VITRO CONSERVATION METHODS

These consist in maintaining the cultures (buds, plantlets derived from nodes or directly from meristems) under physical (environmental factors] or chemical (culture medium composition) stressed conditions that make it possible to extend, as much as possible, the interval of transference into the fresh media, without affecting the viability of the cultures. The methods to reduce the in vitro cultivated plant growth include the reduction of temperature and light during storage, the incorporation of growth retardants in the medium, and the induction of osmotic stress in the medium, or a combination of all these.

Temperature

Temperature reduction has been the most used way to curb culture growth. Most of the in vitro cultures are maintained at temperatures between 12 and 200C: at lower temperatures, the growth rate decreases but the reduction depends on the species.

Nutrients Concentration

The reduction of the carbohydrate concentration and the nitrogenous components of the nutritive medium may affect the growth rates. In addition, the continuous absorption of these nutrients during plantet growth will bring a nutritional defficiency, which could produce a premature death of the plants.

Use of Growth Regulators

The abscisic acid (ABA), phosphon-D, maleic hydrazide, and succinic acid are some of the most frequently used growth regulators.

Osmotic Concentration

Growth limitation caused by osmotic concentration is due to the reduction of the water and nutrients absorbed from the medium. For example, at high concentrations, sucrose acts osmotically and it is highly metabolized. No metabolized osmotics, such as Manitol and Sorbitol, are possibly more efective than sucrose in culture growth limitation.

Evaluation of in Vitro Preserved Material

To evaluate material under these conditions, some important facts about in vitro maintenance such as genetic viability and stability should be considered. The viability evaluation of the in vitro cultures must be systematic. In slow growth conditions, when the sub-culture or transfer period extends during months or years, the frequency of the culture evaluation increases. The most important characteristics to be evaluated in the low-growth cultivars storage of apical buds are: contamination, leaf senescence, the number of green sprouts, the number of viable nodes in relation to the stem length, the presence or not of roots, and callus formation.

MAINTENANCE OF ACCESSIONS IN SEED PROGRAMME

In a seed programme, a group of accessions free of virus is maintained for pre-basic seed production: however, most of them are not propagated in the greenhouse so they are maintained in vitro to be used in the future. The continuous propagation of the in vitro plantlets damages the material, mainly if we consider that environmental conditions are not adequate.

The alternative is to maintain the accessions in conservation media. Each accession must be maintained in test tubes with five replications to avoid possible losses. The maintenance conditions have been tested in CIP's potato collection that consists of more than 5,000 accessions, by means of which it is possible to assure the normal recovery of the material after using stress producers.

The plantlets used in each campaign must come from the maintenance phase. Then, they will be propagated in normal media where their growth will be reestablished, and the elected procedures for plantlet multiplication in the greenhouse will continue.Each season must be initiated with this material to start off with strong plants. In addition, the plantlets taken for multiplication must be replaced, trying to maintain 5 tubes per accession in the conservation media. The sub-cultures in the conservation medium are carried out approximately every one or two years. The conservation medium renovation must pass through a previous sub-culture in the propagation media to rejuvenate the explants.

VIRUS ERADICATION THROUGH MERISTEMS CULTURE AND THERMOTHERAPY

If a healthy plant is sown in the field, it is exposed to infections caused by pathogens as nematodes, fungi, bacteria, phytoplasms, virus and viroids, which have a negative effect on yield, and in some cases may kill the plants. However, not all the plant cells may become infected. A group of cells, which are in a continuous non-differentiated multiplication, are virus-free: the meristem.

The in vitro meristems culture, together with growth at high temperatures produces potato plantlets free from viruses in more than 90 per cent of planted meristems. This routine method was established in the International Potato Centre to obtain virus-free plantlets for national and international distribution. The in vitro maintenance of virus-free plants provides the possibility to maintain all the time a bank of healthy and more vigorous plants, and with a more accelerated growth, than the infected ones. In a seed programme, it is essential to initiate this work with virus-free plantlets since this will affect the seed quality and the yield as well. The virus cleaning procedure is too long and expensive: that is why CIP, through its germplasm distribution programme, has provided a list of the principal virus-free potato varieties to all users.

PROCEDURE

- Approximately 18 to 20 plantlets (virus-infected) are propagated in magentas.
- After a growth period of 20 to 25 days, or when the plantlets are 4-5 cm high, they are placed in the thermotherapy chamber. The growth conditions are: 16 hours of light 34°C; 8 hours of darkness 32°C
- The magentas are maintained in the thermotherapy chamber for one month.
- Afterwards, the magentas are taken out of the chamber, and the outside is cleaned with 980/o alcohol. Then they are introduced to the culture room.
- Meristems are obtained as follows:

Cut the apical portion and remove the leaves that cover the meristem (approximately 3 to 4 leaves); the meristem is observed with a prominent leaf primordium. Remove the meristem with part of the leaf primordium; cut only the translucent portion. Use a new knife.

Place the menistem in the culture medium. Be sure that the meristem is in the tube:

- Evaluate meristem growth and transfer them to fresh media if necessary.
- Each meristem that originates a plant is called <'line», which will be labeled according to the accession it belongs to. For example, Yellow Line 1, Yellow Line 2, Yellow Line 3, etc.
- Five tubes with several plants are propagated to evaluate the potato virus PSTVd, PVT, to determine the host range, the morphologic evaluation, and the in vitro maintenance.
- The results of each evaluation are obtained, and the infected material is replaced with clean material.

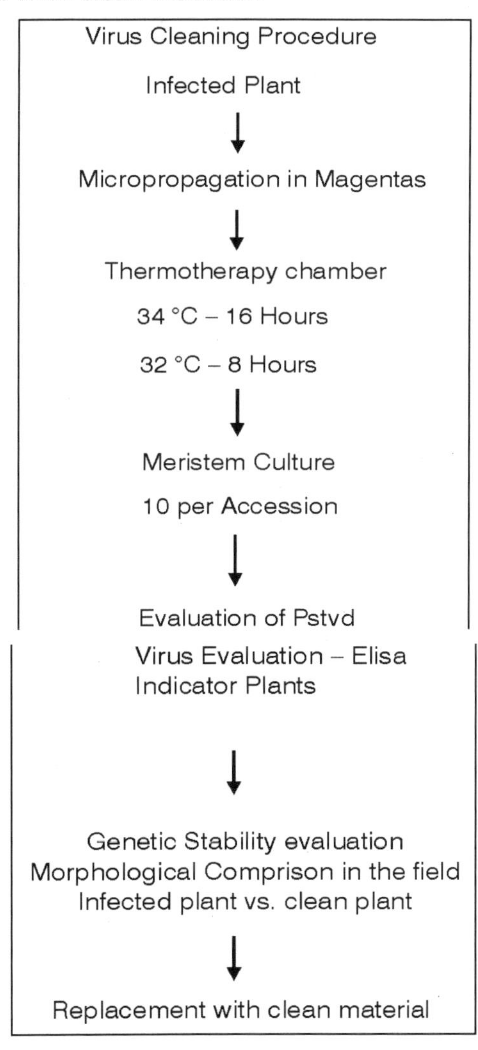

Temperature

The growth of the plantlets at temperatures higher than 20°C is accelerated (Do not use temperatures higher than 30°C).

LOW-COST TISSUE CULTURE TECHNOLOGY

Low-cost tissue culture technology is the adoption of practices and use of equipment to reduce the unit cost of micropropagule and plant production. In many developed countries, conventional tissue culture-based plant propagation is carried out in highly sophisticated facilities that may incorporate stainless steel surfaces, sterile airflow rooms, expensive autoclaves for sterilization of media and instruments, and equally expensive glasshouses with automated control of humidity, temperature and day-length to harden and grow plants.

Many such facilities established at a high cost are high-energy users, and are run like a super-clean hospital. The requirements to establish and operate such tissue culture facilities are expensive, and often are not available in the developing countries. For example, the cost of electricity in the developed countries is much lower, and its supply far better assured than in the developing countries. The same can be said of the supply of culture containers, media, chemicals, equipment and instruments used in micropropagation. Hence, alternatives to expensive inputs and infrastructure have been sought and developed to reduce the costs of plant micropropagation.

ADOPTION OF LOW-COST OPTIONS

Low cost options should lower the cost of production without compromising the quality of the micropropagules and plants. The primary application of micropropagation has been to produce high quality planting material, which in turn leads to increased productivity in agriculture. The generated plants must be vigorous and capable of being successfully transplanted in the field, and must have high field survival. In addition, they should be genetically uniform, free from diseases and viruses, and price competitive to the plants produced through conventional methods. Reducing the cost should not result in high contamination of cultures or give plants with poor field performance.

The foremost requirement of micropropagation is the aseptic culture and multiplication of plant material. Microbe-free conditions need to be maintained in culture containers, and during successive subcultures. In many cases, mistakes in concept or practice can introduce microbes in the culture containers from an external source or the plant material itself (endophytic contamination). As a result, the microbes overgrow the cultures, and wipe them out. Microbes may grow slowly under controlled low temperature, but they proliferate very fast under uncontrolled and high temperature. Thus the adoption of wrong low-cost options may make the production process prone to disasters. Low cost techniques will succeed only if the basic conditions for tissue culture are scrupulously adhered to maintain propagule quality.

Microbial contamination of cultures is known to wipe out work of months, and can turn into a nightmare. The best low-cost option is to discard and dispose of contaminated cultures outright. Avoiding contamination in small R&D laboratories is not a difficult task where only a low number of cultures are handled. However, commercial production involves handling of thousands of cultures each day. It is also essential to maintain such cultures in large numbers under contamination-free conditions, until they are used for either further subculture or hardening and growing-on. Plants do not have an immune system and there is a limitation on the use of antibiotics to circumvent the problem.

Moreover, many of the antibiotics, which are effective against bacteria, fungi, and phytoplasmas, are toxic to plants as well. Obviously, the use of antibiotics is not foolproof or the desired method to rid microbial contamination. Sophisticated state-of-the-art facilities are not a guarantee for prevention of contamination. The laboratories that succeed in an immaculate control of contamination do so by adherence to scrupulous techniques of basic tissue culture. Thus, it is not the sophistication but the procedures that ensure the quality of tissue cultured plants.

QUALITY OF MICROPROPAGULES

Low cost technology means an advanced generation technology, in which cost reduction is achieved by improving process efficiency, and better utilization of resources. Presently, both the developing and the developed countries require low cost technology to progressively reduce the cost of propagule production. In many developing countries, the potential end-users of plants derived from tissue culture have been the resource-rich farmers.

They know the benefits and potential of healthy planting material. Such growers are prepared to risk investment in the high productivity potential of the planting material. For example, hybrid seeds of many vegetables, papaya, rice, and cotton cost 15-20 times more than the price of ordinary varieties.

Yet there is a wide market for them. Hence, the production of low quality plants, just because they are less costly, is not going to be a sustainable approach for the application of micropropagation in agriculture. Lowering of cost of production is possible only if the methods do not compromise the basic imperatives of tissue culture and quality of plants.

IMPORTANCE OF LOW COST TECHNOLOGY

The potential of plant tissue culture in increasing agricultural production and generating rural employment is well recognized by both investors and policy makers in developing countries. However, in many developing countries, the establishment cost of facilities and unit production cost of micropropagated plants is high, and often the return on investment is not in proportion to the potential economic advantages of the technology.

These problems can be addressed by standardizing agronomic practices more precisely (precision agriculture) and by achieving maximum net profits from the crops or by decreasing the unit cost of production or both. The technology is particularly relevant to the propagation of because they command high unit value. However, the market is limited. Over a period of time, many new tissue-culture companies in several developing countries have entered to compete in the limited market. The inevitable result has been the reduction of net margins below viable limits.

Many international organizations also agree that tissue culture technology is very relevant to agriculture, provided the problem of high cost of production is satisfactorily solved. The role of tissue culture was clearly recognized by the FAO in the paper on 'Biotechnology in agriculture, forestry and fisheries - FAO's Policy and Strategy'. The report pointed to the wide use of tissue culture techniques for multiplication of elite clones and elimination of pathogens in planting material. It also pointed to the successful tree regeneration in about 100 forest species and its value for breeding, clonal testing and rapid deployment of superior genotypes.

Several R&D projects have been undertaken to improve the productivity of agricultural, horticultural and forest trees by the European Union under Co-operation in the Field of Scientific and Technical Research (COST). Under this programme, coordinated and funded by the European Union, one of the primary aims has been to reduce micropropagation cost. For example, the objective of 'COST 843' action has been the innovation of low-cost plant propagation methods that enhance sustainable and competitive agriculture and forestry in Europe.

The high costs of labour of micropropagation are a major bottleneck in the EU to fully exploit in vitro culture technology. In the EU, labour currently accounts for 60-70 per cent of the costs of the in vitro produced plants. In another programme, the large-scale production and introduction of bamboo in the EU using tissue culture technology has been undertaken with the main objective of reducing the costs of micropropagation.

FUTURE ROLE

It has been stressed time and again that in the long-term agriculture and forestry need to be sustainable, use little or no crop-protection chemicals, have low energy inputs and yet maintain high yields, while producing high quality material. Biotechnology-assisted plant breeding is an essential step to achieve these goals. Plant tissue culture techniques have a vast potential to produce plants of superior quality, but this potential has been not been fully exploited in the developing countries. During in vitro growth, plants can also be primed for optimal performance after transfer to soil. In most cases, tissue-cultured plants out-perform those propagated conventionally. Thus in vitro culture has a unique role in sustainable and competitive agriculture and forestry, and has been successfully applied in plant breeding, and for the rapid introduction of improved plants. Bringing new improved varieties to market can take several years if the multiplication rate is slow. For example, it may take a lily breeder 15-20 years to produce sufficient numbers of bulbs of a newly bred cultivar before it can be marketed. In vitro propagation can considerably speed up this process.

Plant tissue culture has also become an integral part of plant breeding. For example, the development of pest- and disease-resistant plants through biotechnology depends on a tissue culture based genetic transformation. The improved resistance to diseases and pests enables growers to reduce or eliminate the application of chemicals. The FAO Committee on Agriculture has perceived plant tissue culture as a main technology for the developing countries for the production of disease-free, high-quality planting material, and its commercial applications in floriculture and forestry.

It further points out that tissue culture techniques are being used particularly for large-scale plant multiplication. Micropropagation has proved especially useful in producing high quality, disease-free planting material for a wide range of crops. Tissue cultured based industry also generates much-needed rural employment, particularly for women.In forestry, the availability of tissue culture linked production systems may effectively provide sustainable alternatives to the need for harvesting wood from native forests and natural habitats. Successful protocols now exist for the micropropagation of a large number of forest tree species, and the number of species for which successful use of somatic embryogenesis is increasing.

Thus in the future, it is likely that micropropagation in the forestry sector will become commercially important. Compared to vegetative propagation through cuttings, the high multiplication rates available through micropropagation offer a much quicker capture of genetic gains obtained in forest tree breeding programs. However, the current high costs will also be one of the major impediments to the direct use of micropropagation in many programs.

The broad application of existing technologies to plantation species is important for tree improvement in the tropics. In a small number of plantation programs, micropropagation is being used as an early rapid multiplication step. However, it has been pointed out that the current high costs will be an impediment to the direct use of micropropagules as planting stock. Micropropagation clearly has a role, in the rapid multiplication of the selected clones for conventional production of cuttings.

The direct use of micropropagules as planting stock in industrial plantation can dramatically broaden forestry tree farming if propagation costs are reduced. The availability of micropropagation technologies will also be useful in genetic engineering applications, e.g., the production of plants as a source of "edible" vaccines. There are many other useful plant-derived substances which can be produced in tissue cultures, sometimes more cheaply and reliably than from natural forests and plantations. These include medicinal compounds and drugs now being sought in major prospecting operations in the tropical forests.

Micropropagation has been identified as a suitable technology in the development projects of UNESCO in Africa and the Caribbean; however, the cost of production must be reduced. Practically in all developing countries, the private industry is the most important group that requires cost-effective technology. For example, in India of the 90 commercial micropropagation units established initially, were closed down. Of those engaged in commercial production, many are uneconomic mainly due to the high cost of production and the absence of quality tests. Hence, low-cost tissueculture technology will stay a high priority in agriculture, horticulture, forestry, and floriculture of many developing countries.

TISSUE CULTURE TECHNIQUES

An important aspect of all biotechnology processes is the culture of either the plant cells or animal cells or microorganisms. The cells in culture can be used for recombinant DNA technology, genetic manipulations etc. Plant cell culture is based on the unique property of the cell-totipotency. CELL-TOTIPOTENCY is the ability of the plant cell to regenerate into whole plant. This property of the plant cells has been exploited to regenerate plant cells under the laboratory conditions using artificial nutrient mediums. With the advances made in genetic engineering, it became possible to introduce foreign genes into cell and tissue culture systems. This led to the development of Genetically Modified (Gm) or Transgenic Crops which had improved traits and characteristics.

HISTORY OF CELL CULTURE

In the early 19th century, Schleiden and Schwann proposed the concept of the 'cell theory'. In 1902, Gottlieb Haberlandt, the german botanist and regarded as the father of plant tissue culture, first attempted to cultivate the mechanically isolated plant leaf cells on a simple nutrient medium. He did not succeed in achieving the growth and differentiation of the cultured cells, however, he predicted the concept of growth hormones, the use of embryo sac fluids, the cultivation of artificial embryos from somatic cells, etc.

During the period 1902 - 1930, attempts were made to culture the isolated plant organs such as roots and shoot apices (organ culture). Hanning isolated embryos of some crucifers and successfully grew on mineral salts and sugar solutions. Simon successfully regenerated a bulky callus, buds, roots from a poplar tree on the surface of medium containing IAA which proliferated cell division. Gautheret, White and Nobecourt largely contributed to the developments made in plant tissue culture. White cultured tobacco tumour tissue from the hybrid Nicotiana glauca, and N. Langsdorffii.

The period saw the development of suitable nutrient media to culture plant tissues, embryos, anthers, pollen, cells and protoplasts, and the regeneration of complete plants (in vitro morphogenesis) from cultured tissues and cells. In 1941, van Overbek and co-workers used coconut milk (embryo sac fluid) for embryo development and callus formation in Datura. Steward and Reinert first discovered somatic embryo production in vitro. Maheswari and Guha developed the anther culture for the production of haplid plants. Skoog and Miller advanced the hypothesis of organogenesis in cultured callus by varying the ratio of auxin and cytokinin in the growth medium. Muir developed a successful technique for the culture of single isolated cells wich is commonly known as paper-raft nurse technique (placing a single cell on filter paper kept on an actively growing nurse tissue). In 1952, the Pfizer Inc., New York got the US patent and started producing industrially the secondary metabolites of plants. The first commercial production of a natural product shikonin by cell suspension culture was obtained.

In 1980s using Genetic engineering, for the first time, it was possible to introduce foreign genes into cell and tissue culture systems to develop plants with improved characteristics (transgenic crops) which may contribute to the path towards the second green revolution. A transgenic animal is an individual in which a gene (or genes) from another individual has been artificially inserted with the new genetic information forming a permanent part of the genome and being passed on to their offspring according to the Mendelian laws of inheritance. So far, transgenic cattle, sheep, pigs, rabbits, and chicken have been produced.

Table. List of Transgenic Animals Produced With Their Promoter, Enhancer and Structural Transgenes

Animal	**Genes Tansferred (Promoter or Enhancer/Structural Transgenes)**
Goat	A variant of tPA gene (LAtPA)
Sheep	mMT/hGH, mMT/TK, mMT/bGH, mMT/hGRF, oBLG/hFIX, oBLG/alpha1AT, oMT/oGH
Rabbit	mMT/hGH, hMT/hGH, rbEu/rb,c-myc
Pig	mMT/hGH, mMT/bGH, hMT/pGH, MLV/rGH, MLV/rGH, bPRL/bGH
Fish	hGH, mMT/hGh, mMT/bGal, cd-crystallin SV/hygro, AFP
Cow	BPV, lactoferrin
Chicken	ALV, REV
Mouse	mMT/rGH, mMT/bGH, mMT/oGH, mMT/hGh, mMT/hGRF, mMT/hFIX

After fertilization a mammalian embryo passes through many stages of development such as formation of pronuclei, morula formation, blastocyst formation, implantation etc. Using these early development stages of mammalian embryos, genetic engineers have used different methods to introduce foreign genes into the genome.

METHODS TO INDUCE FOREIGN GENES INTO THE GENOME

Micro-injection in to a Pronucleus

A fertilized egg is held in a pipette by suction and several copies of the foreign genes are injected into one of the pronuclei via a micropipette. Although this has been the most widely used and most successful method, 30 per cent of the treated embryos degenerate and die with in a few hours.

Using Retroviruses as Vectors to Carry Foreign DNA into Morulas

A morula is placed on a culture of fibroblasts infected with retrovirus after removing the zona pellucida layer. The virus cannot penetrate this layer therefore it is essential to get rid of it. The retrovirus, genetically engineered to carry foreign DNA, infect the cells of the morula as they are shed from the fibroblast.

This technique often creates 'mosaic offspring', where some cells contain the foreign gene while others do not. One of the limitations of this technique is that the modified virus cannot leave the transgenic species and under certain conditions it could mutate regaining its ability to cause disease in the tissues where it occurs.

Retroviral Infection of the Stem cells, which are then Injected into the Cavity of Another Blastocyst

In this method, stem cells, from the inner cell mass of an embryo, are infected with genetically-engineered retroviruses, then injected into the central cavity of a different blastocyst. The injected cells colonise the new embryo and participate in the formation of all the tissues, including the ovaries and testes from which cells are formed. This technique has been used to produce transgenic mice and hamsters.

The first transgenic animals produced were mice, into which a gene cloning for rat growth hormone was inserted by microinjection.

Two techniques were used to introduce the new gene into the mouse genome:

- As many as 10000 copies of the rat gene were injected into a pronucleus of a fertilized mouse egg, using a microneedle.
- After treatment with calcium chloride, to make them more permeable, the rat DNA was applied to the outside of a 2-8- celled mouse embryo in culture.

The treated eggs, or embryos were then transplanted into foster mothers. Embryos with the inserted genes were identified by DNA hybridization with small samples of DNA from white blood cells or skin cells. Transformed adult mice produced the growth hormone mainly in their livers rather than pituitary gland where it is normally synthesized.

This technique can be used successfully to transfer other genes such as those affecting the properties of hair, hides, cold tolerance, disease-resistance and milk production which might help to produce strains of farm animals with economic advantage over existing breeds and stocks. In this regard, research is going on to use techniques which can improve the quality of the farm animals. In this regard, following two techniques are of great importance.

IMPROVING THE QUALITY

EMBRYO MANIPULATION

In this method, the fertilized egg is bisected at the two celled stage and each half is transplanted into different regions of the uterus. Using this method the reproductive rate is doubled as we know that female sheep and cattle produce, on an average, one offspring per pregnancy. Hence the farmers can easily increase their farmstock. This technique can also be used to conserve rare breeds. After fertilizing their eggs in the laboratory , the young embryos or rare animals are dissected into anything from 2-8 cells. Each cell, is then transplanted into a surrogate mother of a common breed where it grows to produce a new individual of the rare type.

Another variation of this technique allows surrogate mothers to carry embryos of a different species *e.g.* Horses giving birth to zebras from zebra embryos implanted at the blastocyst stage. The outer layer of zebra embryo is exchanged with the trophectoderm of the horse embryo in order to make it acceptable to the surrogate mother. This exchange of the layer does not affect the development of the inner cell mass which is the true embryo-forming region.

EMBRYO CLONING

Using this technique it is possible to produce many genetically identical copies of an animal. This method has been used to clone mouse however, research is going on to standardize this method to obtain clones of cattle with desirable traits *e.g.* cows with high milk production etc.

Following steps are used in this method:

- An egg from the donor is grown to blastocyst stage under laboratory conditions.
- The egg is dissected to remove the inner cell mass.
- The mass cells are separated into individual cells.
- Injection of a nucleus from each of these cells into a one-celled embryo containing two pronuclei.

- Removal of pronuclei followed by culturing of embryos in the laboratory until the blastocyst stage.
- Transplantation of these embryos into surrogate mothers.

CHIMERAS

Early research into the production of transgenic animals revealed a simple method for producing hybrids, called Chimeras, between closely related species. Sheep-goat chimeras were produced by mixing four celled sheep embryos with eight-celled goat embryos. After removing the zona pellucida from each egg, the eggs were pressed together and incubated at 370C. The cells reorganized and formed hybrid blastocysts. These hybrid blastocysts were then transferred to sheep foster mothers, where they continued their growth and development until the sheep gave birth. Each hybrid offspring contained both sheep and goat cells in all it's tissues which resulted in the coats of these animal having a patchwork kind of pattern with irregular patches of sheep and goat fur.

ENVIRONMENTAL TECHNIQUES

The term "Environment" is defined as our surroundings which includes the abiotic component (the non living) and biotic component (the living) around us. The abiotic environment includes water, air and soil while the biotic environment consists of all living organisms – plants, animals and microorganisms. Environmental pollution broadly refers to the presence of undesirable substances in the environment which are harmful to man and other organisms. In the past decade or two, there has been a significant increase in the levels of environmental pollution mostly due to direct or indirect human activities.

The major sources of environmental pollution are –Industries, Agricultural sources (mainly rural area), anthropogenic sources (man related activities mainly in urban areas), biogenic sources etc. The pollutants are chemical, biological and physical in nature. The Chemical pollutants include- gaseous pollutants (hazardous gases like sulfur dioxide, nitrogen oxide), toxic metals, pesticides, herbicides toxins and carcinogens Etc. The physical pollutants are- heat, sound, radiation, and radioactive substances. The pathogenic organisms and some poisonous and dangerous biological products are the biological pollutants.

Controlling the environmental pollution and the conservation of environment and biodiversity and controlling environmental pollution are the major focus areas of all the countries around the world. In this context the importance and impact of biotechnological approaches and the implications of biotechnology has to be thoroughly evaluated. There have been serious concerns regarding the use of biotechnology products and the impact assessment of these products due to their interaction with the environmental factors.

A lobby of the environmentalists have expressed alarm on the release of genetically engineered organisms in the atmosphere and have stressed on thorough investigation and proper risk assessment of theses organisms before releasing them in to the environment. The effect of the effluents from biotechnological companies is also a cause of concern for everyone. The need of the hour is to have a proper debate on the safety of the use of the biotechnological products. The efforts are not only on to use biotechnology to protect the environment from pollutionbut also to use it to conserve the natural resources. As we all know that microorganisms are known natural scavengers so the microbial preparations (both natural as well as genetically engineered) can be used to clean up the environmental hazards.

DEVELOPMENT OF ALTERNATE CLEANER TECHNOLOGIES USING BIOTECHNOLOGY

Biotechnology is being used to provide alternative cleaner technologies which will help to further reduce the hazardous environmental implications of the traditional technologies. *E.g.* some Fermentation technologies have some serious environmental implications. Various biotechnological processes have been devised in which all nutrients introduced for fermentation are retained in the final product, which ensures high conversion efficiency and low environmental impact.

In paper industry, the pulp bleaching technologies are being replaced by more environmentally friendly technologies involving biotechnology. The pulp processing helps to remove the lignin without damaging valuable cellulosic fibres but the available techniques suffer from the disadvantages of high costs, high energy use and corrosion. A lignin degrading and modifying enzyme (LDM) was isolated from Phanerochaete chrysosporum and was used, which on one hand, helped to reduce the energy costs and corrosion and on the other hand increased the life of the system. This approach helped in reducing the environmental hazards associated with bleach plant effluents.

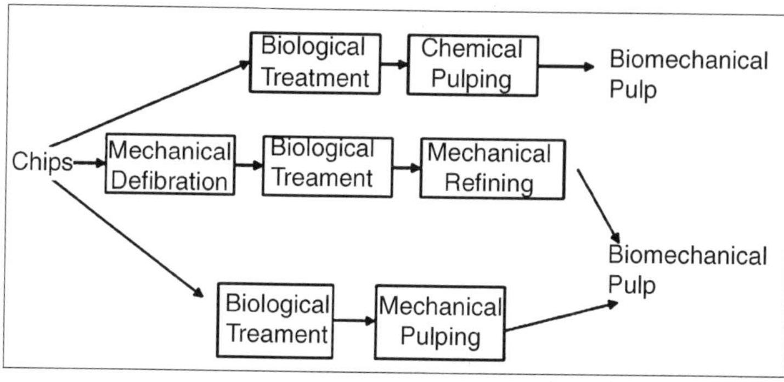

Fig. Integration of Biological Steps in Pulping Process Leading to Lignin Degradation

In Plastic industry, the conventional technologies use oil based raw materials to extract ethylene and propylene which are converted to alkene oxides and then polymerized to form plastics such as polypropylene and polyethylene. There is always the risk of these raw materials escaping into the atmosphere thereby causing pollution. Using biotechnology, more safer raw materials like sugars (glucose) are being used which are enzymatically or through the direct use of microbes converted into alkene oxides.*e.g.* Methylococcus capsulatus has been used for converting alkene into alkene oxides.

BIOREMEDIATION

Bioremediation is defined as 'the process of using microorganisms to remove the environmental pollutants where microbes serve as scavengers. The removal of organic wastes by microbes leads to environmental cleanup. The other names/terms used for bioremediation are biotreatment, bioreclamation, and biorestoration. The term "Xenobiotics" (xenos means foreign) refers to the unnatural, foreign and synthetic chemicals such as pesticides, herbicides, refrigerants, solvents and other organic compounds. The microbial degradation of xenobiotics also helps in reducing the environmental pollution.

Pseudomonas which is a soil microorganism effectively degrades xenobiotics. Different strains of Pseudomonas that are capable of detoxifying more than 100 organic compounds (*e.g.* phenols, biphenyls, organophosphates, naphthalene etc.) have been identified. Some other microbial strains are also known to have the capacity to degrade xenobiotics such as Mycobacterium, Alcaligenes, Norcardia etc.

FACTORS AFFECTING BIODEGRADATION

The factors that affect the biodegradation are: the chemical nature of xenobiotics, the concentration and supply of nutrients and O_2, temperature, pH, redox potential and the capability of the individual microorganism. The chemical nature of xenobiotics is very important because it was found out that the presence of halogens *e.g.* in aromatic compounds inhibits biodegradation.

The water soluble compounds are more easily degradable whereas the presence of cyclic ring structure and the length chains or branches decrease the efficiency of biodegradation. The aliphatic compounds are more easily degraded than the aromatic ones.

BIOSTIMULATION

It is a process by which the microbial activity can be enhanced by increased supply of nutrients or by addition of certain stimulating agents like electron acceptors, surfactants etc.

BIOAUGMENTATION

It is possible to increase biodegradation through manipulation of genes *i.e.* using genetically engineered microorganisms and by using a range of microorganisms in biodegradation reaction.

Depending on the method followed to clean up the environment, the bioremediation is carried out in two ways:

In Situ Bioremediation

In situ bioremediation involves a direct approach for the microbial degradation of xenobiotics at the site of pollution which could be soil, water etc. The adequate amount of essential nutrients is supplied at the site which promotes the microbial growth at the site itself. The in situ bioremediation is generally used for clean up of oil spillages, beaches etc.

There are two types of in situ bioremediation:

1. *Intrinsic bioremediation*: The microorganisms which are used for biodegradation are tested for the natural capability to bring about biodegradation. So the inherent metabolic ability of the microorganisms to degrade certain pollutants is the intrinsic bioremediation. The ability of surface bacteria to degrade a given mixture of pollutants in ground water is dependent on the type and concentration of compounds, electron acceptor and the duration of bacteria exposed to contamination. Therefore, the ability of indigenous bacteria degrading contaminants can be determined I laboratory by using the techniques of plate count and microcosm studies. The conditions of site that favour intrinsic bioremediation are ground water flow throughout the year carbonate minerals to buffer acidity produced during biodegradation, supply of electron acceptors and nutrients for microbial growth and absence of toxic compounds.
2. *Engineered in situ bioremediation*: When the bioremediation process is engineered to increase the metabolic degradation efficiency (of pollutants) it is called engineered in situ bioremediation. This is done by supplying sufficient amount of nutrients and oxygen supply, adding electron acceptors and maintaining optimal temperature and pH. This is done to overcome the slow and limited bioremediation capability of microorganisms.

Advantages of in Situ Bioremediation

- The method ensures minimal exposure to public or site personnels.
- There is limited or minimal disruption to the site of bioremediation.
- Due to these factors it is cost effective.
- The simultaneous treatment of contaminated soil and water is possible.

Disadvantages of in Situ Bioremediation

- The sites are directly exposed to environmental factors like temperature, oxygen supply etc.
- The seasonal variation of microbial activity exists.
- Problematic application of treatment additives like nutrients, surfactants, oxygen etc.
- It is a very tedious and time consuming process.

Ex-situ Bioremediation

In this the waste and the toxic material is collected from the polluted sites and the selected range of microorganisms carry out the bioremediation at designed place. This process is an improved method over the in situ bioremediation method.

On the basis of phases of contaminated materials under treatment ex-situ bioremediation is classified into two:

1. *Solid phase treatment*: This system includes land treatment and soil piles comprising of organic wastes like leaves, animal manures, agricultural wastes, domestic and industrial wastes, sewage sludge, and municipal solid wastes. The traditional clean-up practice involves the informal processing of the organic materials and production of composts which may be used as soil amendment. Composting is a self heating, substrate-dense, managed microbial system which is used to treat large amount of contaminated solid material. Composting can be done in open system *i.e.* land treatment and/or in closed treatment system. The hazardous compounds reported to disappear through composting includes aliphatic and aromatic hydrocarbons and certain halogenated compounds. The possible routes leading to the disappearance of hazardous compounds include volatilization, assimilation, adsorption, polymerization and leaching.
2. *Slurry phase treatment*: This is a triphasic treatment system involving three major components- water, suspended particulate matter and air. Here water serves as suspending medium where nutrients, trace elements, pH adjustment chemicals and desorbed contaminants are dissolved. Suspended particulate matter includes a biologically inert substratum consisting of contaminants and biomass attached to soil matrix or free in suspending medium. The contaminated solid materials, microorganisms and water formulated into slurry are brought within a bioreactor *i.e.* fermenter. Biologically there are three types of slurry-phase bioreactors: aerated lagoons, low shear airlift reactor, and fluidized-bed soil reactor. The first two types are in use of full scale bioremediation, while the third one is in developmental stage.

Advantages of Ex-situ Bioremediation

- As the time required is short, it is a more efficient process.
- It can be controlled in a much better way.
- The process can be improved by enrichment with desired and more efficient microorganisms.

Disadvantages of Ex-situ Bioremediation

- The sites of pollution remain highly disturbed.
- Once the process is complete, the degraded waste disposal becomes a major problem.
- It is a costly process.

Several types of reactions occur during the bioremediation/microbial degradation:

- *Aerobic bioremediation*: When the biodegradation requires oxygen O_2 for the oxidation of organic compounds, it is called aerobic bioremediation. Enzymes like monooxygenases and dioxygenases are involved and act on aliphatic and aromatic compounds.
- *Anaerobic bioremediation*:This does not require oxygen O_2. the degradation process is slow but more cost effective since continuous supply of oxygen is not required.
- *Sequential bioremediation*: Some of the xenobiotic degradation requires both aerobic as well as anaerobic processes which very effectively reduces the toxicity *e.g.* tetrachloromethane and tetrachloroethane undergo sequential degradation.

Use of genetic engineering and genetic manipulations for more efficient bioremediation. In recent years, efforts have been made to create genetically engineered microorganisms (GEMs) to enhance bioremediation. This is done to overcome some of the limitations and problems in bioremediation.

These problems are:

- Sometimes the growth of microorganisms gets inhibited or reduced by the xenobiotics.
- No single naturally occurring microorganisms has the capability of degrading all the xenobiotics present in the environmental pollution.
- The microbial degradation is a very slow process.
- Sometimes certain xenobiotics get adsorbed on to the particulate matter of soil and thus become unavailable for microbial degradation.

As the majority of genes responsible for the synthesis of enzymes with biodegradation capability are located on the plasmids, the genetic manipulations of plasmids can lead to the creation of new strains of bacteria with different degradative pathways. In 1970s, Chakrabarty and his team of co-workers reported the development of a new strain of bacterium Pseudomonas by manipulations of plasmid transfer which they named as "superbug". This superbug had the capability of degrading a number of hydrocarbons of petroleum simultaneouslysuch as camphor, octane, xylene, naphthalene etc. In 1980, United States granted the patent to this superbug making it the first genetically engineered microorganism to be patented.

In certain cases, the process of plasmid transfer was used. *E.g.* The bacterium containing CAM (camphor degrading) plasmid was conjugated with another bacterium with OCT (octane degrading) plasmid. Due to non-compatibility, these plasmids cannot coexist in the same bacterium. However, due to the presence of homologous regions of DNA, recombination occurs between these two plasmids which results in a single CAM-OCT plasmid giving the bacterium the capacity to degrade both camphor as well as octane. A new strain of Pseudomonas sp. (strain ATCC 1915) has been developed for the degradation of vanillate (which is a waste product from paper industry) and sodium dodecyl sulfate (SDS, a compound used in detergents).

BIOTECHNOLOGICAL METHOD TO REDUCE ATMOSPHERIC CARBON DIOXIDE (CO_2)

Carbon dioxide is the gas that is the main cause of green house effect and rise in the atmospheric temperature. During the past 100-150 years, the level of CO_2 has increased about 25 per cent with an increase in the atmospheric temperature by about 0.5 per cent which is a clear indication that CO_2 is closely linked with global warming. There is a steady increase in the CO_2 content due to continuous addition of CO_2 from various sources particularly from industrial processes. It is very clear that the reduction in atmospheric CO_2 concentration assumes significance.

Biotechnological methods have been used to reduce the atmospheric CO_2 content at two levels:

1. *Photosynthesis*: Plants utilize CO_2 during the photosynthesis which reduces the CO_2 content in the atmosphere. The equation for photosynthesis is:
   sunlight
   $$6CO_2 + 6H_2O \rightarrow C_6H_{12}O_6 + 6O_2$$
   Chlorophyll

   The fast growing plants utilize the CO2 more efficiently for photosynthesis. The techniques of micropropagation and synthetic seeds should be used to increase the propagation of such fast growing plants.

   Further, the CO_2 utilization can be increased by enhancing the rate of photosynthesis. The enzyme ribulose biphosphate carboxylase (RUBP-case) is closely linked with CO_2 fixation. The attempts are being made to genetically manipulate this enzyme so that the photosynthetic efficiency is increased.

   Some microalgae like Chlorella pyrenodiosa, Spirulina maxima are known to be more efficient than higher plants in utilizing atmospheric CO_2 for photosynthesis and generate more O_2 than the amount of CO_2 consumed. The growing of these microalgae near the industries and power plants (where the CO_2 emission in to atmosphere is very high) will help in the reduction of polluting effects of CO_2. Using genetic engineering, attempts are going on to develop new strains of these microalgae that can tolerate high concentrations of CO_2. A limited success has already been reported in the mutants ofAnacystis nidulans and Oocystis sp.

2. Biological Calcification- Certain deep sea organisms like corals, green and red algae store CO_2 through a process of biological calcification. As the $CaCO_3$ gets precipitated, more and more atmospheric CO_2 can be utilized for its formation. The process of calcification is as follows:

$$H_2O + CO_2 \rightarrow H_2CO_3$$
$$H_2CO_3 + Ca^{2+} \rightarrow CaCO_3 + CO_2 + H_2O$$

TREATMENT OF SEWAGE USING MICROORGANISMS

The sewage is defined as the waste water resulting from the various human activities, agriculture and industries and mainly contains organic and inorganic compounds, toxic substances, heavy metals and pathogenic organisms. The sewage is treated to get rid of these undesirable substances by subjecting the organic matter to biodegradation by microorganisms. The biodegradation involves the degradation of organic matter to smaller molecules (CO_2, NH_3, PO_4 etc.) and requires constant supply of oxygen. The process of supplying oxygen is expensive, tedious, and requires a lot of expertise and manpower. These problems are overcome by growing microalgae in the ponds and tanks where sewage treatment is carried out. The algae release the O_2 while carrying out the photosynthesis which ensures a continuous supply of oxygen for biodegradation.

PHYSICAL COMPONENTS OF TISSUE CULTURE TECHNOLOGY

The physical components of a typical plant tissue culture facility include equipment and buildings with preparation room, transfer room, culture or growth room, hardening and weaning area, soil-growing area (greenhouses, plastic tunnels), packaging and shipping area, and related facilities – office and store for chemicals, containers and supplies. Careful planning is the first important step when considering the size and location of a facility.

The size of the physical components of a tissue culture facility will vary according to its functional needs – the volume of production. It is recommended that an existing facility should be visited to view the layout and operational needs before starting a new facility. A number of low-cost alternatives can be used to simplify various operations and reduce the costs in a tissue culture facility.

In designing any laboratory, big or small, certain elements in its design and layout are absolutely essential for its successful operation. Correct design of a laboratory will not only reduce contamination, but also achieve a high efficiency in work performance. Properly planned and designed laboratories can reduce both the operational and energy costs. A tissue culture laboratory must be designed to accommodate the equipment and its use in the various stages of micropropagation in the most efficient manner.

GENERAL CONSIDERATION FOR LOCATION

A convenient location for a small laboratory can be a room or part of the basement of a house, a garage, a remodeled office or a room in the header house. The minimum area required for media preparation, transfer and primary growth shelves is about 14 m^2. Walls should be installed to partition different areas.

Before setting up a commercial micropropagation unit, it is essential to check out the area keeping in mind the climate, and access to water, electricity, transportation, and infrastructure for supplies. A temperate climate is usually better suited to tissue culture ventures. This greatly reduces the cost of cooling required to maintain the temperature for optimum growth of the cultures.

The availability of electricity and water is of utmost importance, and should be taken into consideration while choosing the location of the facility. For example in India, of the 76 commercial tissue culture units, nearly 52 are located in and around the cities of Bangalore and Pune, where the climate is moderate.

Hence, However, in cities like Delhi, which have extremes of climate, tissue culture facilities require both heating and cooling. In a facility, which produces five million plants, the electricity cost per thousand plants is around US $0.30 in Bangalore and Pune; the same is about US $0.80 in Delhi. Disruption of power and water supply causes major breakdown in the smooth running of tissue culture units. Poor quality water adds to the cost of media.

Location consideration should also include a check with local authorities about zoning and building permits before construction begins. The buildings should be located away from sources of contamination such as a gravel driveway, parking lot, soil mixing area, shipping dock, pesticide storage, and dust and chemicals from fields. Sanitation of the area is important in selecting location.

The units should be located in areas where insect populations are minimal and air has low dust and pollen counts. In the case of export-oriented units, these should be near an international airport, which will reduce the time lag between packaging and shipment. This is critical to assure timely delivery of quality tissue culture products. For new ventures, the size of the facility should be kept small until the market acceptance is ensured.

DESIGN AND LAY OUT OF THE LABORATORIES

Plant tissue culture laboratories have specific design requirements. Careful initial planning is, therefore, a prerequisite for successful running of a facility. The location and design of the laboratories should take into account isolation from foot traffic, control of contamination from adjacent rooms, thermostatically controlled heating and cooling, water supply and drains for a sink, adequate electrical service, provisions for a fan and intake blower for ventilation, and good lighting. Large sized facilities are frequently built free standing. Although more expensive to build, the added isolation from adjacent activities keeps the laboratory clean. Prefabricated buildings make convenient low-cost laboratories. They are readily available in various sizes in many countries. Prefabricated buildings assembled on site can also be used. A single span building allows for a flexible arrangement of walls for dividing into convenient sized rooms.

The floor should be of concrete or capable of carrying 170.5 kg/m^2 (50 pounds per sq. foot). Walls and ceiling should be insulated to at least R-15, and covered inside with waterresistant material. Windows, if desired, may be placed wherever convenient in the media preparation and glassware washing rooms. The heating system should be capable of maintaining room temperature at 20°C during the coldest part of winter. A minimum of 2cm pipes should be used for water supply. Connection to a septic system or sanitary sewer should be provided. Air conditioning requirements should carefully estimated. Electrical service capacity for equipment, lights and future expansion should be calculated. For safety reasons, the electrical installation should be carried out professionally. Most electrical wiring will require 220 Volts, and autoclaves 230/250 Volts.

The areas such as the media preparation room, inoculation room and growth chambers should be isolated as 'clean zones'. The office, storage area, staff centre and packaging rooms can be maintained under ordinary conditions. The working areas must be demarcated according to the activities involved in the facility. Cleanliness is the major consideration when designing a plant tissue culture laboratory to minimise contamination. A positive pressure module should be installed to circumvent air intake from outside. Routine cleaning and aseptic procedures can decrease contamination losses to less than 1 per cent. An enclosed entrance should precede the laboratories, and sticky mats should be placed to collect dirt from shoes.

The traffic pattern and workflow in the laboratory must be considered to maximise cleanliness. The cleanest rooms or areas are the culture room (aseptic transfer area) and the growth room. There should be no direct access to these rooms from outside. The media preparation area, glassware washing and storage areas should be located away from these rooms. The growth room and aseptic transfer rooms should be fitted with see-through doors and should be adjacent to each other. Traffic through these areas should be minimal and restricted to the personnel working under laminar flow cabinets. Ideally, the media preparation area leads into the sterilisation area, which leads into the aseptic transfer room, and eventually to the growth room. Temperature and fire alarms must be connected directly to telephone lines to give fast warnings. An emergency generator should be available to operate essential equipment during power breakdown.

ESSENTIAL EQUIPMENT

The basic equipment in most tissue culture facilities includes the following:

Autoclave

An autoclave is basically a large-sized but sophisticated pressure cooker, and is used for the sterilisation of the medium, glassware and instruments. Autoclaves of different sizes are available commercially. High-pressure heat is needed to sterilise media, water, and glassware. Certain spores from fungi and bacteria are killed only at 121°C and 1.05kg/sq.cm (15 pounds per sq. inch) pressure. Self-generating steam autoclaves are more dependable and faster to operate.

Laminar Airflow Chamber

The laminar flow chambers provide clean filtered air that allows cultures to be handled under contamination-free environment. Several types of laminar flow chambers are sold on the market and are available in different sizes. The laminar-flow cabinets are located in the culture transfer area. Some large-sized laboratories have sterile rooms in addition to laminar flow cabinets.

Other equipment

The sterilisation of instruments, such as the forceps, scalpel holders and blades is achieved with either gas flamed burners or with glass-bead sterilizers. The medium preparation room usually has the following equipment. A refrigerator-freezer to store chemicals and stock solutions, weighing scales for large amounts of over 10 g, and an analytical balance with1 mg accuracy, a magnetic stirrer for the agitation, and a pH meter. Small laboratories may locate the refrigerator under the workbench to save space. High quality balances are essential in a tissue culture laboratory. Most laboratories have top loading balances, which allow quick and efficient weighing. A hot plate with an automatic stirrer is needed to for preparing the media before autoclaving.

The pH meter is needed to determine pH of the media. Some laboratories use pH indicator paper, however this method is considerably less accurate, and can severely affect the results. An aspirator can be attached to a water tap for filter sterilisation of chemicals and for surface sterilisation of the plant material. However, vacuum pumps are faster and more efficient, but also more expensive. A drying oven is required to keep glassware such as beakers, flasks and cylinders, and is also useful for dry sterilisation of scalpels and glassware, such as Petri dishes, pipettes and others. The media containing carbon sources (e.g. sugars) and growth regulators are sterilised in the autoclave, but sometimes, aseptic filtration is better to avoid breakdown of heat-labile chemicals. The water still is also located in the medium preparation area. To prepare media, distilled or de-ionised water is generally used, although tap water can be used in some cases.

OPTIONAL EQUIPMENT

A variety of non-essential equipment is used in tissue culture laboratories. The specific requirements determine what need to be purchased. Microwave ovens are convenient for defrosting stock solutions and pre-heating agar media. Most laboratories have a dissecting microscope to excise small explants. Laboratory glassware washers or regular dishwashers can be used for replacing manual labour. Automatic media dispensers are helpful to pipette pre-set volume of media. A gyratory shaker or a reciprocal shaker is necessary if micropropagation is based on liquid media or suspension cultures. Computers, photocopiers and fax machines are helpful for easy data management and maintenance of records. Some of the equipment may be costly, but goes a long way in saving time and labour and is essential for rapid communication in the competitive world.

ACTIVITY SPECIFIC REQUIREMENTS

Based on the different activities of a tissue culture, a facility can be divided into semiclean, clean and ultra-clean areas. The semi-clean areas comprise of the washing room, office and staff restrooms, where there is no need for maintaining sterile conditions. The clean areas encompass the media preparation and sterilisation rooms, which have to be sufficiently clean. High sterility has to be maintained in the culture transfer rooms and the growth rooms, which constitute the ultra-clean areas.

Glassware Washing and Storage Area

The glassware washing area should be located near the sterilisation and medium preparation rooms. This area should have at least one large sink but two sinks are preferable. Adequate workspace is required on each sides of the sink; this space is used for glassware soaking and drainage. Plastic netting can be placed on surfaces near the sink to reduce glassware breakage and enhance water drainage. The outlet pipe from the sink should be of PVC to resist damage from acids and alkalis. Both hot and cold water should be available and the water still and de-ionisation unit should be located nearby. The choice of electrical washers should be based on the projected use, durability, reliability and cost, and service availability. In India and some other developing countries, where labour is relatively cheap, washing is done manually. The washing room should be swapped periodically.

Mobile drying racks can be used and lined with cheesecloth to prevent water dripping and loss of small objects. Ovens or hot air-cabinets should be located close to the glassware washing and storage area. Dust-proof cabinets and storage containers should be installed to allow for easy access to glassware. When culture vessels are removed from the growth area, they are often autoclaved to kill contaminants and to soften semi-solid media. It should be possible to move the vessels easily to the washing area. The glassware storage area should be close to the wash area to expedite storage and access for media preparation.

The Store

It is advisable to have a separate area for storage of chemicals, apparatus and equipment. It would not only facilitate constant availability but also save cost from bulk purchase. Chemicals required in small amounts should not be purchased in large quantities as they may lose their activity, pick up moisture or get contaminated. Such problems can be overcome by purchasing small lots on a regular basis.

Media Preparation and Sterilisation Area

The media preparation room should have smooth walls and floors, which enable easy cleaning to maintain a high degree of cleanliness. Minimum number of doors and windows should be provided in this room but within the local fire safety regulations. This reduces cost and contamination. The media preparation and sterilisation can be carried out in the same area but preferably in different rooms, which need not be separated with doors. Media preparation area should be equipped with both tap and purified water. An appropriate system for water purification must be selected and fitted after careful consideration of the cost and quality.

A number of electrical appliances are required for media preparation; hence, it is essential to have safety devices like fire extinguisher, fire blanket and a first aid kit in the media preparation room. A variety of glassware, plastic ware and stainless steel apparatus is required for measuring, mixing, and media storage. These should be stored in the cabinets built under the worktables and taken out for use as and when required.

This would save the cost and space for building storage shelves. The use of glassware should be kept at a minimum, as it will help in reducing losses due to breakage. As far as possible, plastic ware and stainless steel vessels should be used, as they are much cheaper and more durable than glassware. The water source and glassware storage area should be in or near the medium preparation area. Work bench tops, suitable for comfortable working while standing should be 85 to 90 cm high and 60 cm deep. The workbench tops should be made with plastic laminate surfaces that can tolerate frequent cleaning.

Sterilisation Room

The sterilising room should be in continuation with the media preparation room. The layout must be planned in such a way that it ensures the smooth movement of the containers from the washing to the media preparation and sterilisation room. The sterilisation room must have walls and floors that can withstand moisture, heat and steam. An exhaust should be fitted to remove the warm and moist air. The exhaust fan should have an outer cover to prevent entry of outside air.

The fan cover should open only when the fan is in operation. In small tissue culture facilities, costly autoclaves can be replaced by simple pressure cookers. However, for large volume media making, horizontal or vertical autoclaves should be installed. Double door autoclaves, which open directly into the media storage room, may be costly but reduce contamination. A cheaper alternative is to transfer the sterilised media to the adjoining room through a hatch window.

Transfer room

The most important work area is the culture transfer room where the core activity takes place. The transfer area needs to be as clean as possible and be a separate room with minimal air disturbance. Walls and floors of the transfer room must be smooth to ensure frequent cleaning. The doors and windows should be minimal to prevent contamination, but within local safety code. There is no special lighting requirement in the transfer room. The illumination of the laminar airflow chamber is sufficient for work. Sterilisation of the instruments can be done with glass-bead sterilizers or flaming after dipping in alcohol, usually ethanol.

The culture containers should be stacked on mobile carts (trolleys) to facilitate easy movement from the medium storage room to the transfer room, and finally to the growth room. The chair seats of the transfer operators should be comfortable, as they have to work for long periods in the same place. Fire extinguishers and first aid kits should be provided in the transfer room as a safety measure. The personnel should leave shoes outside the room. Special laboratory shoes and coats should be worn in this area. Ultraviolet (UV) lights are sometimes installed in transfer areas to disinfect the room; these lights should be used only when people and plant material are not in the room. Safety switches can be installed to turn off the UV lights when regular room lights are turned on.

Growth Room

Growth room is an equally important area where plant cultures are maintained under controlled environmental conditions to achieve optimal growth. It is advisable to have more than one growth room to provide varied culture conditions since different plant species may have different requirements of light and temperature during in vitro culture. Also, in the event of the failure of cooling or lighting in one room, the plant cultures can be moved to another room to prevent loss of cultures.

In the growth room, the number of doors should be minimal to prevent contamination. There is no need for windows in the growth room, except when natural light is used. When artificial lighting is used, the external light can interfere with the photoperiod and temperature of the growth room. Depending on the amount of available space and cost, the culture containers can be placed on either fixed or mobile shelves. Mobile shelves have the advantage of providing access to cultures from both sides of the shelves. The height of the shelves should not exceed 2m. High shelving requires step-up stools to place and remove cultures, being dangerous and time consuming. The primary source of illumination in the growth room is normally from the lights mounted on the shelves.

Overhead light sources can be minimised, as they would be in use only while working during the dark cycle. Plant cultures may not receive uniform light from the conventional downward illumination. Lights directly fitted to the racks create uneven heat distribution. This leads to high humidity within the culture containers, which in turn can cause hyperhydricity. Sideways illumination is an alternative, which requires less number of lights, and provides more uniform lighting. But care has to be taken not to break the lights while moving the cultures across the shelves.

CONTROL OF GROWING CONDITIONS

Controlled temperature, lighting and relative humidity, and shelving need to be considered in planning the growth room. These vary depending on the size of the growth room, its location, and the type of plants cultured. For example, a small growth room located in the cool North American climate can be placed in an unheated or minimally heated basement. The chokes (ballasts) of the fluorescent lights need not to be separated; rather they can serve as a source of heat. Excess heat can be dissipated from the growth room, and used for heating other areas in the basement. In such a situation, solid wooden shelves with space between shelves can be used and prevent culture vessels on shelf above the lights from becoming over-heated. However, a large growth room located above ground needs to have the light chokes installed outside the room. Shelves in large growth room can be of glass or metal wire mesh.

Temperature Control

Temperature is a primary concern in growth rooms; it affects decisions on installation of lights, control of relative humidity, and type of shelving. Temperature in the growth room is usually controlled with air conditioners. Generally, temperatures are kept around 22°C. Heating is provided from conventional heating systems and can be supplemented with heat from light chokes.

In most developing countries, cooling the growth room is usually a bigger problem than heating. Cooling can be provided with heat pumps, air conditioners and exhaust fans. Using open windows to cool culture rooms leads to contamination during summer and humidity problems in winter.

Lighting Control

Some plant cultures can be kept in complete darkness; however, most culture rooms need to be illuminated at 1 Klux [134.5 μ mole/m2/s (microeinsteins per second per sq. centimetre or approximately 1076 foot candle) with some up to 5 to 10 Klux (672-1345 μ mole/m^2/s). The plant species and/or propagation scheduling determines the light intensity. The developmental stage of the plants also determines if wide spectrum or cool white-fluorescent lights are to be used. Rooting is strongly influenced positively with far-red light; therefore, wide spectrum lights should be used during Stage III and cool-white lights during Stage I and II.

Automatic timers are needed to maintain the desired photoperiod. Reflectors can be placed over bulbs to direct the light downwards and evenly. Heat generated by lights may cause condensation and temperature problems. Small fans placed at the end of the shelves increase airflow and decrease heat build-up. Reflective glossy paint on the walls provides an even light distribution as well as reduces the number of lights required. Relative humidity (RH) is difficult to control inside the culture vessels, but wide fluctuations in the growth room have a deleterious effect. Cultures can dry out if the room's RH is less than 50 per cent. Humidifiers can be used to correct this problem. If the RH becomes too high, a dehumidifier is recommended.

Shelving

Shelves within the growth rooms vary depending upon the situation and the plants grown. Frames for the shelves can be made from 1.25cm (half-inch) thick angle iron. Shelves built from rigid wire mesh to allow maximum air movement and minimise shading should be used. Wood is inexpensive to build shelves. The wood for shelving should have smooth exterior, and should be painted white to reflect light.

Expanded metal is more expensive than wood, but provides better air circulation. Tempered glass is sometimes used for shelves to increase light penetration, but it is more prone to breaking. Air spaces of 5 to 10cm between the lights and shelves decrease heat on upper shelves and reduce condensation in culture vessels. A room that is 2.4m high will accommodate 5 shelves, each 45cm. apart, when the bottom shelf is 10 cm above the floor.

GREENHOUSE FACILITY

A critical stage in plant tissue culture is the interim phase between the laboratory and field conditions. In vitro derived plants need to be gradually hardened to field conditions. Plant hardening is usually carried out under greenhouse that ensures high survival of the tissue-cultured plants in the field. There are three types of greenhouses: Ground to ground, Gable, and Quonset type. The most commonly used greenhouse is the Quonset type. It contains movable or fixed benches with hardening tunnels on them. The size of the greenhouse must be based on the scale of production. Greenhouse glazing can be of glass or fibreglass.

Polyethylene films or sheets of polycarbonate or acrylic can also be used. Air inflated double polyethylene covering is the most economic. Appropriate light, shading and blackout systems can be achieved with supplementary lighting. Drip irrigation systems, misting and fogging can be installed as needed. Greenhouses erected in warm climates should have fan-assisted drip pad cooling especially during summer. Greenhouses in colder climates need to be heated. Floor and bench systems can be used for heating and cooling the air. Low cost plastic pipes can be used to circulate warm air, which are adequate and cost effective.

PACKAGING AND SHIPPING

A separate area should be designated for packaging in a commercial tissue culture unit. Packaging materials such as cardboard cartons and labels should be stored in this area. The type of packaging of a particular plant depends greatly on the temperature zones through which the consignment has to pass from the point of shipment to its destination.

For example, if the consignment of tissue-cultured plants is passing anywhere within temperate zones, it is only necessary to take care of the frost conditions and pack accordingly. But if the consignment is passing from a tropical country to a temperate country, it becomes necessary to take care of the temperature zones in different places through which the consignment passes. Thus, the type of packaging of tissue-cultured products varies with plant and destination. For example, in a package of 22 kg, about 40,000 lily plantlets can be packed.

However, in the same parcel, only 20,000 Spathiphyllum plantlets or 10,000 Syngonium plantlets can be packed because of the higher respiration rates in these species. The heat build-up is much more in Spathiphyllum and Syngonium plants than the dormant lilies. Similarly, only 8000 Gerbera or 7000 Cordyline plants can be packed in the same package. Cooling material such as ice, dry ice and other material take up much of the packaging space.

Before loading and shipping, the packed items should be properly counted and rechecked. The cartons containing the cultures for shipping to the customer should be properly labelled with the names and addresses of the consignor and the consignee, and the details of the commodity (storage temperature, handling, etc.). Adequate care must be taken while packaging large consignments, so that there is no disturbance or damage during transit.

To prevent any sort of delay, ensure that the consignment is accompanied by documents such as invoice, packing list, import permit, phytosanitary certificate and Generalized System of Preferences (GSP).

COST OF FACILITY

The building of a tissue culture facility includes the cost of land, construction, electrical installation and plumbing. If the available funds are limited, well-designed laboratories can be established by modifying existing structures. For example, a two or three room house or a trailer with appropriate modifications can be converted into a medium sized micropropagation facility. Depending on the production and storage capacity, a facility can be small- less 100,000 plants, medium- 100,000 to 500,000 plants, or large-scale- 500,000 to 2,000,000 plants per annum.

The initial investment required for setting up a low cost, medium scale tissue culture laboratory (ca. 195 m^2) in India. Land is sold at a premium, especially near large cities. In India, the land and building costs work out to about US $62500. A house, ca. 195 m^2, can be converted into a tissue culture facility with a capacity of 200,000 plants. The rental charges for such a structure would be around US $5000 per year. The initial investment would thus include only plumbing and electrical work, reducing the capital cost to US $2750. Thus, in large cities, it would be more economic to rent than build the premises. If the hardened tissue-cultured plants were the end products, an additional investment of US $20,000 would be required to erect a greenhouse. If the production is limited to-in-agar and ex-agar products (rooted or non-rooted micro-cuttings), greenhouse would not be required.

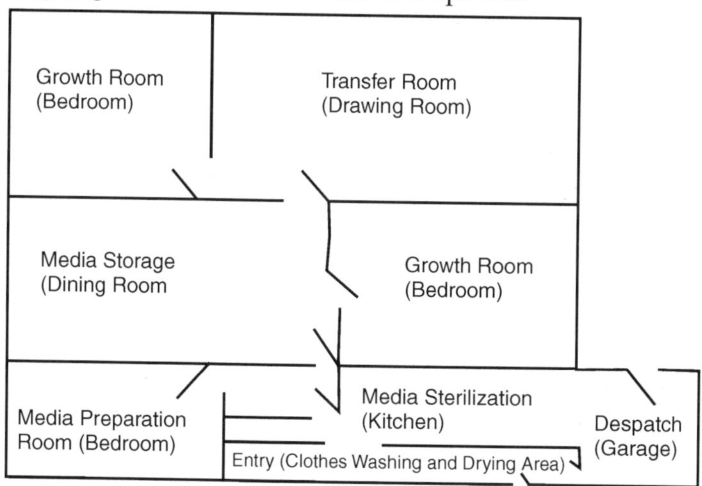

Fig. Conversion of a House into a Micropropagation Facility

The basic equipment will cost US $14000. The consumables, such as glassware, chemicals and disposable items would require another US $4200.

Besides the fixed capital, a commercial tissue culture unit has the recurring expenses of rent, building maintenance, electricity and overheads. The estimated cost of furniture that includes racks with illumination source for growth room, cupboards, benching and chairs would be US $7400. Other miscellaneous items such as labelling and sealing machines would cost US $50.

STRATEGIES FOR COST REDUCTION

For a commercial tissue culture unit to be successful, it is essential to constantly find means to increase the efficiency of production, and bring down the cost of production. A number of low-cost alternatives can be used to simplify various operations and reduce costs in a tissue culture facility.

Contamination Control

The loss of cultures increases the cost of production. Contamination in cultures is caused from natural contaminants such as dust, air-borne particles, bacterial and fungal spores, fibres, and hair. Man-made contamination occurs mainly from body, clothing, and from faulty procedures in the laboratory.

It has been estimated that each operator generates a minimum of 1 to 5 million particles (bigger than 0.5 μ M diameter) per minute. Modern, high quality clean rooms have environment control systems that minimise contamination. Wearing laboratory coats (which are mandatory in Class 100 clean rooms) reduces contamination from clothing, skin and hair. Maintaining cleanliness in other working areas is equally important.

There is continuous entry of contaminants with the worker and through air entering the growth room. A set routine procedure should be followed to reduce contamination in the laboratory; however, such procedures will vary with the location. In the absence of high cost systems, careful planning and appropriate design of the laboratory goes a long way in reducing contamination.

Washing and Sterilising Operations

Costly dishwashing machines for cleaning the culture containers should be replaced by manual washing if labour is relatively inexpensive. The washed culture containers can be dried in sun instead of costly hot air ovens. In small-scale laboratories, the autoclaves can be replaced by large sized pressure cookers, which are much cheaper. Instead of having one or two huge horizontal autoclaves, which generate hot air pockets in the sterilising rooms, it is better to have more of smaller vertical autoclaves, which keep the air cool. It is a normal practice to use costly aluminium foil for wrapping instruments for sterilisation. This can be replaced by stainless steel containers, which are autoclavable. Savings can also be made in media and containers.

Transfer Operations

Instead of Petri dishes as stage for manipulation of culture, stainless steel plates, ceramic tiles and brown wrapping paper can be used; all of these can be autoclaved. Ethanol for hand and workbench sterilization can be replaced by industrial spirit. The careless handling of inflammables used for sterilization can be hazardous. Glass bead sterilizers can be used to sterilize forceps and scalpels instead of the conventional flame sterilization using spirit lamps or gas cylinders. Commercial bleach has also been successfully used in several laboratories for bench and instrument sterilization to reduce cost, and to prevent fire hazard.

Culture Maintenance in Growth Rooms

Plant cultures can be maintained in rooms with air conditioners and tube lights instead of highly priced plant growth chambers. The conventional method of downward illumination can be replaced by sidewise lighting systems, which not only reduces the number of lights but also provides more uniform illumination to the cultures. In the tropical and Mediterranean regions, the electrical lighting systems can be replaced by sunlight. The tissue culture laboratories in Cuba produce millions of tissue-cultured sugarcane, pineapple and banana plants using natural light.

Production Planning

Most plants are seasonal in demand. To meet the requirements of extremely large number of plants, commercial production has to be backed by well defined working procedures and monitoring performance of the operation. Commercial laboratories should produce a range of plants for different seasons to maximise the use of the facilities throughout the year. This lowers the unit cost of plant production. Large-scale micropropagation is a labour-intensive process, and therefore organising the availability of personnel is quite important. To reduce organisational problems in big units, the management structure must be well planned.

Supporting information systems such as inventory control, production scheduling, space utilisation and daily targets should be well defined. To produce quality products on large-scale, there should be good co-ordination among the technicians, supporting staff, supervisors and the researchers. The job description and reporting system should be very clearly stated. In addition, personnel selection and training is critical for successful large-scale production. The production needs to be periodically reviewed to meet needs of the customers.

Scale of Operation

The physical components required for the functioning of a commercial tissue culture laboratory can be either scaled up or down according to the interests of the propagator. The correct design of a laboratory, big or small, will help maintain asepsis, thereby increasing the efficiency of the unit and achieving a high standard of work. A micropropagation company in India converted a three-room apartment into a medium-sized tissue culture laboratory. Plant such as Spathiphyllum, Syngonium, Ficus, hosta, calla lily, gerbera, and cordyline, were produced on a commercial scale.

Delivery of the tissue-cultured plants to the tune of 2.5 million per year has been made from this unit by adopting various low cost alternatives. Similar tissue culture facilities, aptly called 'bio-factories', in villages of Cuba produce up to one million banana plants and 2.5 million sugarcane plants annually. Significant cost reduction for large-scale production can be achieved with automation and mechanisation. Although, automation of certain steps of micropropagation has been investigated for the past 20 years, its commercial use has not been

PLANT TISSUE CULTURE

Plant tissue culture refers to growing and multiplication of cells, tissues and organs of plants on defined solid or liquid media under aseptic and controlled environment. The commercial technology is primarily based on micropropagation, in which rapid proliferation is achieved from tiny stem cuttings, axillary buds, and to a limited extent from somatic embryos, cell clumps in suspension cultures and bioreactors.

The cultured cells and tissue can take several pathways. The pathways that lead to the production of true-to-type plants in large numbers are the preferred ones for commercial multiplication. The process of micropropagation is usually divided into several stages i.e., prepropagation, initiation of explants, subculture of explants for proliferation, shooting and rooting, and hardening. These stages are universally applicable in large-scale multiplication of plants. The delivery of hardened small micropropagated plants to growers and market also requires extra care.

Plant tissue culture refers to growing and multiplication of cells, tissues and organs on defined solid or liquid media under aseptic and controlled environment. Plant tissue culture technology is being widely used for large-scale plant multiplication. The commercial technology is primarily based on micropropagation, in which rapid proliferation is achieved from tiny stem cuttings, axillary buds, and to a limited extent from somatic embryos, cell clumps in suspension cultures and bioreactors.

EXPLANT SOURCE

Plant tissue cultures are initiated from tiny pieces, called explants, taken from any part of a plant. Practically all parts of a plant have been used successfully as a source of explants. In practice, the "explant" is removed surgically, surface sterilized and placed on a nutrient medium to initiate the mother culture, that is multiplied repeatedly by subculture. The following plant parts are extensively used in commercial micropropagation.

Shoot-tip and meristem-tip culture: Shoots develop from a small group of cells known as shoot apical meristem. The apical meristem maintains itself, gives rise to new tissues and organs, and communicates signals to the rest of the plant. Shoot-tips and meristem-tips are perhaps the most popular source of explants to initiate tissue cultures. The shoot apex explant measures between 100 to 500 μ m and includes the apical meristem with 1 to 3 leaf primordia. The apical meristem of a shoot is the portion lying distal to the youngest leaf primordium, and is ca.100 μ??m in diameter and 250μ?m in length with 800-1200 cells. In practice, shoot-tip explants between 100 to 1000 μ m are cultured to free plants from viruses. Even explants larger than 1000 μ m have been frequently used. The term "meristem-tip culture" has been suggested to distinguish the large explants from those used in conventional propagation.

Nodal or axillary bud culture: This consists of a piece of stem with axillary bud culture with or without a portion of shoot. When only the axillary bud is taken, it is designated as "axillary bud" culture. Floral meristem and bud culture: Such explants are not commonly used in commercial propagation, but floral meristems and buds can generate complete plants. Other sources of explants: In some plants, leaf discs, intercalary meristems from nodes, small pieces of stems, immature zygotic embryos and nucellus have also been used as explants to initiate cultures.

Cell suspension and callus cultures: Plant parts such as leaf discs, intercalary meristems, - stem-pieces, immature embryos, anthers, pollen, microspores and ovules have been cultured to initiate callus. A callus is a mass of unorganized cells, which in many cases, upon transfer to suitable medium, is capable of giving rise to shoot-buds and somatic embryos, which then form complete plants. Such calli on culture in liquid media on shakers are used for initiating cell suspensions. Liquid suspension cultures maintained on mechanical shakers achieve fast and excellent multiplication rates. However, in commercial micropropagation, calli are cultured mostly in bottles and flasks kept on semi-solid or liquid media. To a limited extent, bioreactors have become popular for somatic embryogenic cultures. It is considered that some day robotics could be adapted to bioreactorbased micropropagation.

PATHWAYS OF CULTURED CELLS AND TISSUES

The cultured cells and tissue can take several pathways to produce a complete plant. Among these, the pathways that lead to the production of true-to-type plants in large numbers are the popular and preferred ones for commercial multiplication. The following terms have been used to describe various pathways of cells and tissue in culture.

Regeneration and Organogenesis

In this pathway, groups of cells of the apical meristem in the shoot apex, axillary buds, root tips, and floral buds are stimulated to differentiate and grow into shoots and ultimately into complete plants. In many cases, the axillary buds formed in the culture undergo repetitive proliferation, and produce large number of tiny plants. The plants are then separated from each other and rooted either in the next stages of micropropagation or in vivo (in trays, small pots or beds in glasshouse or plastic tunnel under relatively high humidity). The explants cultured on relatively high amounts of auxin (e.g. (2,4-D, 2,4-dichlorophenoxyacetic acid) form an unorganized mass of cells, called callus. The callus can be further sub-cultured and multiplied.

The callus shaken in a liquid medium produces cell suspension, which can be subcultured and multiplied into more liquid cultures. The cell suspensions form cell clumps, which eventually form calli and give rise to plants through organogenesis or somatic embryogenesis. In some cases, explants e.g. leaf-discs and epidermal tissue can also generate plants by direct organogenesis and somatic embryogenesis without intervening callus formation, e.g. in orchardgrass. Dactylis glomerata L.. In organogenesis the cultured plant cells and cell clumps (callus) and mature differentiated cells (microspores, ovules) and tissues (leaf discs, inter-nodal segments) are induced to differentiate into complete plants to form shoot buds and eventually shoots, and rooted to form complete plants.

Somatic Embryogenesis

In this pathway, cells or callus cultures on solid media or in suspension cultures form embryo-like structures called somatic embryos, which on germination produce complete plants. The primary somatic embryos are also capable of producing more embryos through secondary somatic embryogenesis. Although, somatic embryogenesis has been demonstrated in a very large number of plants and trees, the use of somatic embryos in large-scale commercial production has been restricted to only a few plants, such as carrot, date palm, and a few forest trees.

Somatic embryos are produced as adventitious structures directly on explants of zygotic embryos, from callus and suspension cultures. Somatic embryos and synthetic seeds (embryos encapsulated in artificial endosperm) hold potential for large-scale clonal propagation of superior genotypes of heterogeneous plants. They have also been used in commercial plant production and for the multiplication of parental genotypes in large-scale hybrid seed production. In many species, somatic embryos are morphologically similar to the zygotic embryos, although some biochemical, physiological and anatomical differences have been documented.

The synthetic auxin, 2,4-D is commonly used for embryo induction. In many angiosperms, e.g., carrot and alfalfa, subculture of cells from 2,4-D containing medium to auxin-free medium is sufficient to induce somatic embryogenesis. The process can be enhanced with the application of osmotic stress, manipulation of medium nutrients, and reducing humidity. Selection of embryogenic cell lines has also been successfully used. For example, selection for unique morphotypes in grapevine cultures allows production of high quality embryos with predicable frequency. A major problem in large-scale production of somatic embryos is culture synchronization. This is achieved through selecting cells or pre-embryonic cell clusters of certain size, and manipulation of light and temperature, temporary starvation or by adding cell cycle synchronizing chemicals to the medium. Cytokinins seem to play a key role in cell cycle synchronization and embryo induction, proliferation and differentiation. Abscisic acid is crucial in all the stages of somatic development, maturation and hardening.

Synthetic Seeds

The concept of production and utilization of synthetic seeds (somatic embryo as substitutes for true seeds) was first suggested by Murashige in 1977. Synthetic seeds can be produced either as coated or non-coated, desiccated somatic embryos or as embryos encapsulated in hydrated gel (usually calcium alginate).

Successful utilization of synthetic seeds as propagules of choice requires an efficient and reproducible production system and a high percentage of post-planting conversion into vigorous plants. Artificial coats and gel capsules containing nutrients, pesticides and beneficial organisms have long been thought as substitutes for seed coat and endosperm. However, this technology is still in the developmental stage, and currently cannot compete with the other methods of commercial plant propagation.

Process of Micropropagation

The process of plant micro-propagation aims to produce clones (true copies of a plant in large numbers). The process is usually divided into the following stages:

- *Stage* 0: Pre-propagation step or selection and pre-treatment of suitable plants.
- *Stage I*: Initiation of explants - surface sterilization, establishment of mother explants.
- *Stage II*: Subculture for multiplication/proliferation of explants.
- *Stage III*: Shooting and rooting of the explants.
- *Stage IV*: Weaning/hardening.

These stages are universally applicable in large-scale multiplication of plants. The individual plant species, varieties and clones require specific modification of the growth media, weaning and hardening conditions.

A rule of the thumb is to propagate plants under conditions as natural or similar to those in which the plants will be ultimately grown ex-vitro. For example, if a chrysanthemum variety is to be grown under long day-length for flower production, it is better to multiply the material under long-day length at stages III and IV. There is a wide option to undertake production of plant material up to a limited number of stages. For example, many commercial tissue culture companies undertake production up to Stage III, and leave the remaining stages to others.

Pre-Propagation Stage

The pre-propagation stage (also called stage 0) requires proper maintenance of the mother plants in the greenhouse under disease- and insect-free conditions with minimal dust. Clean enclosed areas, glasshouses, plastic tunnels, and net-covered tunnels, provide high quality explant source plants with minimal infection. Collection of plant material for clonal propagation should be done after appropriate pretreatment of the mother plants with fungicides and pesticides to minimize contamination in the in vitro cultures. This improves growth and multiplication rates of in vitro cultures.

The control of contamination begins with the pretreatment of the donor plants. They may be prescreened for diseases, isolated and treated to reduce contamination. The explants are then brought to the production facility, surface sterilized and introduced into culture. They may at this stage be treated with antibiotics and fungicides as well as anti-microbial formulations, such as PPM. The explants are then culture indexed for contamination by standard microbiological techniques, which are occasionally supplemented with tests based on molecular biology or other techniques.

Stage I

This stage refers to the inoculation of the explants on sterile medium to initiate aseptic culture. Initiation of explants is the very first step in micropropagation. A good clean explant, once established in an aseptic condition, can be multiplied several times; hence, explant initiation in an aseptic condition should be regarded as a critical step in micropropagation. More than often, explants fail to establish and grow, not due to the lack of a suitable medium but because of contamination. The explants are transferred to in vitro environment, free from microbial contaminants.

The process requires excision of tiny plant pieces and their surface sterilization with chemicals such as sodium hypochlorite, ethyl alcohol and repeated washing with sterile distilled water before and after treatment with chemicals. After a short period of culture, usually 3 to 5 days, the contaminated explants are discarded.

The surviving explants showing growth are maintained and used for further subculture. In herbaceous plants e.g. potato, chrysanthemum, carnation, streptocarpus, strawberry, and African violet; the explant sources are meristems, apical- and axillary buds, young seedlings, developing young leaves and petioles, and unopened floral buds. The following low cost options can be adapted to initiate explants:

Sterile Instrument Technique

This method assumes that most of the deep-seated meristems and those covered by leaves or other integuments (e.g. floral bracts) are sterile. In this procedure, the explant is washed with sterile water, rinsed in ethanol, and instruments are sterilized every time they touch the surface of the explant, and the explant is moved to a new location on the dissection stage.

Surface Sterilization Technique

This is by far the most commonly used method. The explants are washed in sterile water, rinsed in ethanol, and surface sterilization is achieved by using chemicals with chlorine base. Calcium or sodium hypochlorite based solutions, 1–3 per cent (v/v) are usually used for soft herbaceous materials. A cheap and ready-made sterilant is 5-7 per cent solution of 'Domestos'- a toilet disinfectant which contains 10.5 per cent v/v sodium hypochlorite, 0.3 per cent sodium carbonate, 10.0 per cent sodium chloride and 0.5 per cent (w/v) sodium hydroxide and a patented thickener). The explants are washed in sterile distilled water before and after sterilization. Other surface sterilants used include mercuric chloride (avoid its use as far as possible, since it is highly toxic), hydrogen peroxide, and potassium permanganate.

For Soft Tissues

- Wash explants from perennial plants for 1-2 hr in tap water. Eliminate this step for material from glasshouse grown plants.
- Wash in sterile distilled water three to four times for 5 to 10 minutes each.
- Dip in 95 per cent ethanol for 3 to 5 seconds.
- Wash once again with sterile distilled water for 5 minutes.
- Surface-sterilize in 5 per cent 'Domestos' (v/v) for 20-25 minutes.
- Wash with sterile distilled water three times for 10 minutes each.
- Drain water droplets by placing on pre-sterilized blotting paper.
- Transfer explants singly to the medium.

N.B.: Sterilize forceps each time to transfer explants to avoid cross-contamination.

For woody stems (e.g. roses, hardy shrubs, and trees):

- Collect stems, shoots, buds and store at 5 oC till needed.
- Rinse in ethanol for 3 to 5 seconds.
- Rinse in 1- per cent sodium hypochlorite (20 per cent bleach) for 10 minutes.
- Place lower parts of stems in flasks in 2 per cent sucrose and 200 PPM 8-hydroxyquinoline citrate at 23+2 oC. For items collected in September/October, add 50-PPM GA3.

After that l0 PPM GA will help break the dormancy.

- Re-cut the bottom of stem and replace the solution after 2 days.

- Excise the softwood from the developed shoots and use material for explants or for rooting.
- Surface-sterilize as in the above protocol.

Do not forget to sterilize forceps and scalpel every time for the transfer of explants to fresh solutions. Use sterile containers in the protocol of surface sterilization. If explants become brown or pale at the end of the protocol, reduce the strength of 'Domestos' to 2.5 per cent. Alternatively, dip explants in 10 per cent 'Domestos' for 2 minutes and then proceed to surface sterilize with 3-5 per cent 'Domestos' for 20 minutes. If basal contamination is observed after 2-3 days of culture, explants can sometimes be rescued by removing the basal end by making a single cut with a sharp scalpel and re-culturing on fresh medium.

Stage II

Stage II is the propagation phase in which the explants are cultured on the appropriate media for multiplication of shoots. The primary goal is to achieve propagation without losing the genetic stability. Repeated culture of axillary and adventitious shoots, cutting with nodes, somatic embryos and other organs from Stage I leads to multiplication of propagules in large numbers.

The propagules produced at this stage can be further used for multiplication by their repeated culture. Sometimes it is necessary to subculture the in vitro derived shoots onto different media for elongation.

Stage III

The in vitro shoots obtained at Stage II are rooted to produce complete plants. If the proliferated material consists of bud-like structures (e.g. orchids) or clumps of shoots (banana, pineapple), they should be separated after rooting and not before.

Many plants (e.g. banana, pineapple, roses, potato, chrysanthemum, strawberry, mint, several grasses and many more) can be rooted on half-strength-MS medium without any growth-regulators. Good sturdy well-rooted plants are essential for high survival during weaning and later transfer to soil. This stage is labour intensive and expensive. The process of in vitro rooting has been estimated to account for approximately 35–75 per cent of the total cost of production. Efforts should be made to combine rooting and acclimatization stages.

Stage IV

At this stage, the in vitro micropropagated plants are weaned and hardened. This is the final stage of the tissue culture operation after which the micropropagated plantlets are ready for transfer to the greenhouse. Steps are taken to grow individual plantlets capable of carrying out photosynthesis.

The hardening of the tissue-cultured plantlets is done gradually from high to low humidity and from low light intensity to high intensity conditions. If grown on solid medium, most of the agar can be removed gently by rinsing with water.

Plants can be left in shade for 3 to 6 days where diffused natural light conditions them to the new environment. The plants are then transferred to an appropriate substrate (sand, peat, compost, etc.), and gradually hardened. Low-cost options include the use of plastic domes or tunnels, which reduces the natural light intensity and maintains high relative humidity during the hardening process. If the plants are still joined together after rooting, these should be planted as bunches in the soil and separated after 6 to 8 weeks of growth.

DELIVERY TO THE GROWERS

The delivery of the rooted and hardened small micropropagated plants to growers and market requires extra care. In some cases, plant losses can occur during shipment and handling by growers. This is particularly true when the plants are not fully hardened and rooted or not grown for sufficient duration after transfer to soil. Growers should be given clear instructions how to handle the material provided. Apart from the economic loss, poor survival of planted material erodes the confidence of growers in the technology. The transfer of individual plants to soil in black plastic or polythene bags is widely used as a low-cost option to provide fully-grown banana plants directly to farmers in many developing countries.

5

Methods and Technology of Plant Tissue Culture

APPLICATIONS OF PLANT TISSUE CULTURE

Plant tissue culture has wide applications in agriculture. One major advantage of plant tissue culture is the production of disease and/or pest resistant varieties, thus indirectly increasing the crop yield. Commercially, it is used directly for the propagation of plants that are hard to propagate in natural conditions. Horticultural plants such as orchids, roses, banana, strawberries, potatoes, apples, etc. are successfully cultured in in-vitro conditions. Plants with valuable secondary products are grown in the controlled conditions by using plant tissue culture technique. It is also used for the propagation of medicinal herbs on a large scale. With plant tissue culture, it is possible to generate virus-free plantlet of vegetatively propagated plants.

In experimental biology such as plant breeding, cell biology, biotechnology and genetics, plant tissue culture is applied in order to solve plant related problems. It allows screening of cells for the desirable characters such as early fruit bearing, disease resistance and drought resistance. Another important application is generation of a novel hybrid by crossing two distantly related species having advantageous traits. In-vitro fertilization and/or pollination of plants is possible, irrespective of the hindrances in natural conditions. Overall, plant tissue culture is used for the conservation of germplasm.

THEORY OF TISSUE CULTURE

The theoretical basis for plant tissue culture was proposed by Gottlieb Haberlandt, German Academy of science in 1902 on his experiments on the culture of single cell. The first true cultures were obtained by Gautheret from cambial tissue of Acer pseudoplatanus. The term plant tissue culture (Micro propagation) is generally used for the aseptic culture of cells, tissues, organs and their components under defined chemical and physical conditions *in vitro*.

The basic concept of the plant body can be dissected into smaller part termed as "explants" and any explants can be developed into a whole plant. It is a central innovative areas of applied plant science, including agriculture and plant biotechnology.

This technique is effective because almost all the plants cell are totipotent; In each cell possesses the genetic information and cellular machinery necessary to generate the whole organism. Since, this technique can be used to produce a higher number of plants that are genetically similar to a parent plant as well as to another. Two concepts, plasticity and totipotency, are the central processes to understand the regeneration and plant cell culture. Plants, due to its longer life span and sessile nature, have developed a greater ability to overcome the extreme conditions.

Most of the processes include in plant development and the growth, adapt to environmental conditions. When the plant cells and tissues are cultured *in vitro*, most of them are generally exhibit a very high degree of plasticity, which allows one type of organ or tissue to be initiated from another type. Like this way, the whole plant can be subsequently regenerated. These maintenance of genetic potential is called totipotency. The plant tissue culture medium is an artificial nutrient supplement of organic and inorganic nutrients used for cultivation of plant tissue media. The appropriate composition of the medium largely determines the success of the culture. The culture media used for the *in vitro* cultivation of the plant cells are composed of three basic components.

- Essential elements (normal ions) supplied as a complex mixture of salts.
- An organic supplements providing vitamins and amino acids.
- A source of fixed carbon which is usually supplied as sucrose.

When cultured in an appropriate medium having auxin and cytokinin, explants will give rise to an unorganised, growing and dividing mass of cells called callus. Callus cultures are initiated from a small part of an organ or tissue segment called the explants on a growth supporting solidified nutrient medium under sterile conditions.

Any part of the plant organ or tissues may be used as the explants. At the time of callus formation, there is some degree of dedifferentiation happens both in morphology and metabolism. One of the major consequences of this dedifferentiation is that most plant cultures lose their ability to perform photosynthesis. The necessitates of the addition of other components such as carbon and vitamins source to the culture media, in addition to the unusual mineral nutrients.

MORPHOLOGY OF CALLUS

Callus varies considerably in appearance and texture, ranging from hard nodular cell masses to friable soft ones. They maybe white or creamish, orange, green either in whole or part as a result of chloroplast development. The shape of individual cells within the callus mass ranges from the near spherical or markedly elongated. A typical unorganised plant callus initiated from a new explants or piece of previously initiated calli has three stages of development.

1. The induction of cell division.
2. A period of active cell division during which differentiated cells lose specialised features they may have acquired and become dedifferentiated. Cell division usually occurs in the outer layer of the explants.
3. Period when cell division slows down on ceases and when within the callus, there is increasing cellular differentiate.

Callus culturing is performed in the dark while light can be encourage the differentiation of the callus. At the time of long term culture, the culture may loss the requirement for cytokinin and auxins. Manipulation of the auxins to cytokinin ratio in the medium can leads to the development of shoots, roots or somatic embryos from which the plant can be subsequently produced.

Callus culture is useful for many purposes:

- Callus is the starting material for the suspension culture which cells are separated.
- It helps in the production of secondary plant products.
- It is useful for the synthesis of starting compounds that are subsequently modified to yield the desired product.
- It is the starting materials for vegetative propagation of plants.

Based on the availability of the various *invitro* techniques, the dramatic increase in their application to various problems in basic biology, agriculture, horticulture, and forestry.

The applications can divide conveniently into six broad areas:

1. Cell behaviour.
2. Plant modification.
3. Germplasm storage and pathogen –free plants.
4. Clonal propagation.
5. Product formation.
6. Improved varieties.

STRUCTURE AND PARTS OF PLANT CELL CULTURE

Plant cells like animal cells are eukaryotic cells i.e. they contain membrane bound nuclei and cell organelles. Plant cell differs from the animal cell in having three distinctive structures – cell wall, vacuoles and plastids. Plant cells lack centrioles and intermediate filaments, which are present in animal cells.

PARTS OF A PLANT CELL STRUCTURE

Plant cells are classified into three viz. parenchyma cells, collenchyma cells and sclerenchyma cells based on the structure and function. Now let us see the different parts of a plant cell.

Cell Wall

Cell wall is the outermost rigid layer composed of cellulose, hemicellulose, pectin and sometimes lignin. The function of cell wall is protection, structural support and also it helps in filtering mechanism.

Cell Membrane

Cell membrane also called as plasma membrane is present inside the cell wall and surrounds the cytoplasm. It connects the intracellular components with the extracellular environment and helps in protection and transportation.

Plasmodesmata

Plasmodesmatas are small openings, which connect plant cells with each other enabling transport and communication between them.

Vacuole

Vacuoles are large membrane bound compartments, which stores compounds and provides storage, excretory and secretory functions. The membrane surrounding vacuole is called tonoplast.

Cytoplasm

Cytoplasm is filled up by cytosol, which is a gelatinous, semitransparent fluid.

Nucleus

Nucleus is a specialized organelle, which contains the plant's hereditary material i.e. DNA (Deoxyribonucleic Acid). It also contains structures, which regulates the cell cycle, growth, protein synthesis and reproductive function.

Plastid

Plastids are organelles responsible for the photosynthetic activity and for the manufacture and storage of chemical compounds in plants. Chloroplast is an important form of plastid containing chlorophyll pigment, which helps in harvesting light energy and converting it to chemical energy.

Mitochondria

Mitochondria are oblong shaped organelles that are also known as "the powerhouse of the cell". They are responsible for breaking down the complex carbohydrate and sugar molecules to simpler forms that the plants can use.

Endoplasmic Reticulum

Endoplasmic reticulum is an organelle responsible for the manufacturing and storage of chemical compounds like glycogen and steroids, translation and transportation of protein. It is also connected to the nuclear membrane so as to make a channel between the cytoplasm and the nucleus.

Golgi Apparatus

Golgi apparatus also known as golgi complex is an organelle responsible for the processing and packaging of macromolecules such as proteins and fats, which are synthesized by the cell and prepares them for transportation.

Ribosome

Ribosomes are organelles, which are made up of 60 per cent RNA (Ribonucleic Acid) and 40 per cent protein and play an important role in protein translation.

Microbodies

Microbodies are single membrane bound organelles, globular in shape and contains degradative enzymes. Most common microbodies are peroxisomes.

Microtubules

Icrotubules are straight, hollow, tubular cylinders, which make up the cytoskeleton. They are responsible for structural support and transport of the cell.

Microfilaments

Microfilaments are thin filaments of the cytoskeleton and are responsible for structural support of the cell.

MODEL AND PARTS OF PLANT CELL CULTURE

Plant cells differ from animal cells in that they have three different structures known as cell wall, vacuoles and plastids. However, they lack centrioles and intermediate filaments which are present in animal cells. Read more on plant cell structure and parts.

Cell Membrane

It permits the waste material to exit the cell and regulates the movement of materials in and out of the cell. It also acts as a barrier between the inside of the cell and the outside, so that the chemical conditions on both sides can be different.

Cell Wall

It provides rigid structural support to the cell from which structures, like, leaves and stems are produced. It allows the circulation and distribution of water, minerals and nutrients in and outside the cell. The wall also controls the presence of pathogen microbes and development of tissues in the cell.

Golgi Body

It stores, packages, and circulates the lipids and proteins made in the endoplasmic reticulum. It stores the proteins into packages called vesicles.

Rough Endoplasmic Reticulum

It synthesizes and exports proteins and glycoproteins throughout the cell.

Lysosomes

These organelles digest food by breaking down larger molecules into smaller ones, using special proteins.

Cytoplasm

It helps in maintaining the cell shape. Along with this, it plays an important role in the internal movement of cell organelles, cell mobility and muscle fibre contraction. It also distributes oxygen and nutrients to different parts of the cell.

Nucleolus

It is the most prominent structure in the nucleus wherein ribosomes are made.

Vacuole

The plant cell contains a large, single vacuole (an enclosed compartment) which is used to store water, compounds and minerals that help in plant growth.

Ribosomes

These are packets of RNA (Ribonucleic Acid) and are called as protein builders or synthesizers of the cell.

Chloroplasts

They are the food producers of the cell, containing chlorophyll, the green pigment essential for photosynthesis. Their main function is to generate sugars and starches.

Nucleus

It's the most important part of the cell, comprising chromosomes, i.e. structures made up of genetic information that helps in cell growth and reproduction. Read more on cell nucleus: structure and functions.

Nuclear Envelope

It's an enclosure that surrounds nucleus and its contents. Unlike cell membrane which has pores and spaces for RNA and proteins to pass through, it keeps the chromatin and nucleolus inside the nucleus.

Smooth Endoplasmic Reticulum

It's main function is to package proteins for transport, synthesize membrane phosolipids, and secrete calcium. It also performs transformation of bile pigments, glycogenolysis (the breakdown of glycogen), and detoxification of different drugs and chemical agents.

Mitochondria

It provides energy to the cell by combining sugar molecules with oxygen to generate carbon dioxide and water.

Amylosplast

It's an organelle present in some plant cells that stores starch.

Druse Crystal

It's a granular type of crystal found in plant vacuoles. It is composed of calcium oxalate and is considered to deter herbivory.

Centrosome

Also known as Microtubule Organizing Centre, it's a region in the cell where microtubules are produced that perform a variety of functions ranging from transportation to structural support.

Though both plant and animal cell centrosomes play similar roles during cell division, plant cell centrosomes are simpler and do not contain centrioles.

Peroxisomes

These are membrane bound packets of oxidative enzymes that convert fatty acids into sugar and assist chloroplasts in photo-respiration.

Golgi Vesicles

These vesicles transfer proteins and lipids to different parts of the cells.

PLANT CELL MODEL

Following are some simple instructions or plant cell model ideas, to construct a model of a plant cell:

- Enumerate and collect all possible materials required for the model and look out for the different cell parts in a biology textbook.
- Use a heavy sturdy wooden base to construct the base of the model so that it can hold the weight and facilitate presentation of the model.
- Construct the cell nucleus and ensure that it's round in shape. Paint the nucleus in one colour so that it can be identified as a distinct cell part.

- With the help of a dowel rod or thick, straight stick, connect the nucleus to the base of the model. For this, drill a hole into the wooden base, stick the dowel rod into the hole and connect the nucleus to the rod.
- The cell wall and cell membrane of the plant cell model should be rectangular in shape. The outer wall is the cell wall which should be made of a strong material like hard plastic or wood and the cell membrane can be made of cellophane or thin plastic. Now attach both the walls to the nucleus and the model base.
- Add various other key components of the plant cell to the model. Like, for chlorophyll, use a piece of green fabric stuffed with cotton. Sew it and attach it to the model.
- Finally, prepare a report on the various plant cell parts and their corresponding functions.

PLANTS HARDENING IN TISSUE CULTURE

Tissue culture raised plants are subjected to specific culture regimes aimed at making them capable of surviving the uncontrolled and harsher *ex vitro* environment.

The process of acclimatisation of the *in vitro* grown plants to the normal environment is hardening. The plants may be transferred to mist chamber or green house and then to partial shade and to filed conditions slowly.

This will make the plant to adapt to a newer environment. Proper care must be taken during hardening and minimum mortality should be seen. According to the micro climate available in the area the hardening process vary from plant to plant.

Hardening mixtures may be:

- Vermiculite
- Perlite
- Cocopeat
- Peat soil
- Compressed peat pellets
- Soil
- Sand

Vermiculite:

- It has excellent property of improving soil aeration
- Permanent, clean, odourless, non-toxic and sterile with pH 7.0-9.5.
- Vermiculite is very light and easy to handle, it easily mixes with soil, peat, composted pine bark and other composted organic materials, and fertilizers.

Perlite:

- Perlite is a unique volcanic mineral

- The surface of each particle is covered with tiny cavities which provide an extremely large surface
- These surfaces hold moisture and nutrients and make them available to plant roots
- Because perlite is sterile, it is free of disease, seeds, and insects, a component of growing mixes with materials such as peat moss or bark.

Cocopeat:

- Cocopeat is the 'coir fibre pith' or 'coir dust' produced as a bi-product when coconut husks are processed.
- Coco peat products feature increased water use efficiency, superior air capacity, and work well with both organic and inorganic plant nutrients.

Fired clay pebbles:

- The clay is formed into pellets and fired in rotary kilns at 1200°C.
- Expanded fired clay pebbles are light in weight, do not compact, and are completely reusable, completely inert, pH neutral, and do not contain any nutrients.

Hardening mixture ratios:

- Soilrite mix: 75 per cent pat moss, 25 per cent expanded perlite. pH 5-6.5.
- Peatmoss: perlite: vermiculite (1:1:1).
- Red soil: coirpith:sand (2:1:1)
- Peatmoss: sand: vermiculite (1:1:1).
- Peatmoss: perlite (2:1).

Mist Chamber

This is a structure wherein plants root after taken out from the *in vitro* media. Mist camber is provided with high humidity and temperature that induced faster rooting. Proper sanitation should be followed for preventing the plants from pests and diseases. Fog formation also induces rooting and acclimatisation. Foliar application of fertilizers is also possible by spraying as droplets now a day. Automated systems with timer and temperature control are also available.

Green House

It is the structure where plants can be hardened after rooting in the *in vitro* medium or for rooting and acclimatisation. This structure provides an aseptic environment to the tissue culture or any plants that are propagated.

Location:

- The building should be in South (or) south West direction in order to be cool.
- Normally no trees are allowed to grow over the green house because it may damage the structure if the branches fall. Trees may be grown

so as to cover the upper surface to give a cool effect during summer and allow the light to pass during winter seasons.

- Good drainage facilities should be provided because it may invite pests and diseases during hardening

TYPES OF GREEN HOUSES

Green houses can be built according to the space available and the taste of the administrator.

Structural Materials

Frames: Wood, Galvanised steel or Aluminum may be used as framing material and the frames may be covered with glass, ridged fibre glass or plastic fibre.

Frames Types:

- Quonset.
- Gothic
- Rigid-frame
- 'A' Frame
- Post and rafter.

Coverings

Glass- These are used traditionally which has a pleasing appearance and inexpensive to maintain part from its high degree of permanency.

Disadvantage- Easily broke, initially expensive to build and a good foundation is required.

Fibre glass:

- Light wight, strong, hail proof.
- Use clear, transparent/translucent grades.
- Resin covering allows dirt to be retained by exposed fibres

ouble wall plastic sheets of acrylic/Poly carbonate gives long life:

- Heat saving covers
- Having two layers of ridged plastic separately
- Retain more heat.

Film plastic:

- Replacing frequently than others.
- Low cost
- Inexpensive.
- Made of polyethelene(PE), Polyviny lchloride(PVC), Copolymess.
- PE-UV inhibitors.

Foundation and Floors

Poured concrete flooring is not recommended because it may stay wet and slippery from soil mix media. Floor should be covered by several inches of gravel- good drainage and produce humidity.

ENVIRONMENTAL SYSTEM

Heating

To maintain desired day/night temperature in winter:

- A 220 V circuit electric heater, small gas heaters.
- Paint contains black to attract heat and fill them with water to retain its.
- Distribute by forced hot air, radiant heat, hot water, stream

Air Circulation

- To maintain temperature uniformly.
- Warm air rises to top and cool air settles at bottom.
- Place the fans diagonally opposite corners.
- Develop circular (oval) pattern of air movement.
- Operate-Winter and Turn off-Summer

Ventilation

- Exchange of inside air for outside air to control temperature and remove moisture/ replenish CO_2.
- Natural ventilation-roof vents on ridge line with side inlet vents.
- Mechanical ventilation-exhaust fan to move air out one end of the GH while outside air enters the other end through motorised inlet louvers

Cooling

The air temperature decreases evaporative cooling:

- Evaporative cooler -works when the RH of out side is low.
- Commercial Green houses-places the pads on the air inlets at one ends of the Green house uses the exhaust fans at the other end of GH to pull the air through the houses.

Controllers/Automation

Thermostats- to control individual units/central controller with one temperature sensor.

Watering System:-Hand watering is acceptable:

- Automatic watering-time clock/mechanical evaporative sensor.
- Mist spray-humidity maintaining CO_2 and Light:
- Electric lighting + CO_2 injection-increase the yields.
- Bottled CO_2, dry ice, and combustion of sulphur free fuels- CO_2 sources.

POLY HOUSE TECHNOLOGY

Timber end frames: -single span poly houses.

Cladding for poly houses:

- Standard films (White/Clear films)
- Advanced films (Polytherm+Luminance)
- Standard visqueen polythene180mμ(720 g) UV inhibited films.
- Clear film transmission -89 per cent
- White film -70 per cent.
- Advanced visqueen UV inhibited film-anti condense and heat retentive property.

Nicotrap: light transmission -48 per cent

Energy saving system:

- Advanced Polythene films.
- Air inflated double skinned polythene.
- Internal/External thermal screen system.

Twin Cladding:

- Offers 50 per cent energy savings.
- Two sheets of films.
- Inner sheet-Tight.
- Outer sheet-slake.

Ventilation:

- Vital growth &Diseases control.
- Simple laurel vents to sophisticated mounted roll-up side vents on multi span.
- Height of vents 1-2m(48 ft)-normally.

Role up side vents:

- Works by film being wound around a tube.
- Turned manually by a small gear box unit.

Ground rails:

- Fitted to make of tunnel provide to foundation tubes
- For fixing roof sheet at ground level-Aluminium box base rail.

Cross bracing:

- Horizontal and vertical bracing arrangement across the span of hoop.
- Increase resistance to snow and wind loading.

Net House Technology

This structure is mostly used in cultivation of tissue culture plants like ornamentals, cutflowers like gerbera, carnation, chrysanthemum, Roses, etc. Here netted structures are used wherein nets are used and free flow of air is possible. The nets are having more life and rooting is successful in propagated plants.

Quality Maintenance Inside the Green House

Certain guidelines have been given by the National certification system for tissue culture raised plants by the Department of Biotechnology, GOI are as follows:

- Facility for ventilation to control excess of RH during monsoon
- Excessive watering of plants to be avoided

- Plants should be monitored regularly for their growth and presence of any disease or pest
- Dead plants should be removed immediately to avoid any possible attack of saprophytic fungi
- Fungal infestation in Green house particularly during monsoon season is very common. If present the plants should be sprayed with suitablc fungicides
- Wherever possible use of compost at the Green house stage should be avoided because that may invite contamination
- Any kind of treatment given to the plant such as fertilizers, fungicides, pesticides, etc. must be recorded for reference just in case something wrong with the plants
- All mortalities taking place in the green house/poly house should be arrive at the transplantation losses.

APPROACH TO FUMIGATION APPLICATION

A new approach to fungicide application in green house has been named as Envirosol technology. This technology uses carbon dioxide to deliver pesticides as aerosol droplets into encklosed spaces like green houses. Envirosol products such as Permigas, Pestigas and Insect gas are commercially available. For instances Floragas and Hortigas have been developed as post harvest fumigants for the treatment of cut flowers and asparagus (Carpenter and Stocker, 1992).

Growers use high volume spraying with thermal pulse jet foggers in controlling green house diseases and insect pests. These applications too have run-off of excessive pesticides. Hence, specific user friendly green house fungicides application technologies to be developed in future.

HARDENING TRANSPLANTS

The transplanting process can be a shock to rapidly growing seedlings especially when set out into the cold windy garden in the spring. This is especially true for transplants started in the greenhouse, cold frame, hotbed or home. These young seedlings can be made somewhat resistant to heat, cold temperatures, drying and whipping winds, certain types of insect injury, injury from blowing sand and soil particles and low soil moisture by a process termed "hardening".

The term "hardening" refers to any treatment that results in a firming or hardening of plant tissue. Such a treatment reduces the growth rate, thickens the cuticle and waxy layers, reduces the percentage of freezable water in the plant and often results in a pink colour in stems, leaf veins and petioles. Such plants often have smaller and darker green leaves than non-hardened plants. Hardening results in an increased level of carbohydrates in the plant permitting a more rapid root development than occurs in non-hardened plants.

Cool-season flower and vegetable plants can develop hardiness allowing them to withstand subfreezing temperatures. Unhardened cabbage seedlings have been reported to be damaged by temperatures of –2°C (28°F) while hardened cabbage will tolerate temperatures as low as –6°C (22°F). Warm-season types of plants even when hardened, will not withstand temperatures much below freezing. If transplanted to the garden or field prior to the average last killing spring frost, such plants should be provided protection by hot caps or other such devices.

Method

Any of the following can be used to harden transplants.
A combination of all these techniques at one time is more effective:

- Gradually reduce water - water lightly at less frequent intervals but do not allow the plants to wilt severely.
- Expose plants to lower temperature than is reported as optimal for their growth. If biennials are exposed to cold for an extended period, they may bolt in lieu of developing properly. Note: Placing the plants outside during the day to encourage hardening and then bringing the plants back into the warm house during the night often reverses the hardening process. Plants could be placed in a cold frame or other area that does not freeze during the night hours without lose of the hardening process.
- Do not fertilize, particularly with nitrogen immediately before or during the hardening process. A starter solution or liquid fertilizer could however be applied to the hardened transplants one or two days prior to transplanting into the garden or at the time of transplanting.
- Gradually expose the plants to more sunlight. This results in the development of a thicker cuticle layer thereby reducing water loss.

Cautions

Hardening is not necessary for all transplants. Splittstoesser recommends that with the exception of tomatoes, plants that are susceptible to frost should not be hardened. Overly hardened plants while withstanding unfavourable outside conditions are slow to get started and may never overcome the stress placed on the plant during the hardening process. Lorenz and Maynard recommend that plants be hardened for no longer than seven to ten days before planting to the garden site.

EFFECTS OF HARDENING

The young seedlings get used to the outdoor environment throughout the hardening process. Indoor growing conditions provide the proper amount of nutrients, a relatively stable temperature, and a steady supply of moisture and light. Once the plants are mature enough for outdoor growing, you can damage the unsuspecting plants if you lift them from their perfect life and dump them into the harsh landscape. The stems are soft due to the water content, and are easily damaged by winds, debris and pests. Hardening allows the plants to toughen up gradually in the new environment and build up their strength.

Hardening Process

The recommended time for the entire hardening process is two to three weeks. The strawberry plants should have at least three true leaves before the start of the hardening process.The first step takes place indoors with a reduction of water. The soil should remain moist but dry out more between watering. The stems have a chance to firm up and thicken as water in the stems is replaced with fibre.

Lower the temperature by 10 or 15 degrees to prepare for the cooler outdoor temperatures. Maintain this change in the environment for about a week. Watch for signs of stress, such as wilting or loss of leaves. Fertilizer is not applied during the hardening off period so the plant does not produce new growth. The new growth is susceptible to disease and pests that could damage the entire plant.

Temporary Outdoor Location

The outdoor location for hardening off should be in the shade and away from strong winds. Direct or strong sunlight causes the seedlings to dehydrate quickly and could result in plant failure within an hour. The plants stay outdoors for about an hour the first couple days, and exposure time increases each day until the plants remain outside without damage for the entire day. Once the plants show acclimation to the daytime temperatures, move the pots into a sunny location and start acclimation to the direct sunlight.

The first day of direct sunlight is for an hour and the time gradually increases until the plant shows no signs of wilting or damage when outdoors for the entire day. If the temperature is too hot in the direct sun, move the plants outdoors during the evening and bring them back indoors before nightfall.

Transplanting the Strawberries

The strawberry plants are acclimatised when the seedlings spend the entire day and night outdoors without damage. Once this occurs, the plants should be transplanted into the permanent growing site when all danger of frost is over. Transplant on a cloudy day or late in the evening to avoid stress. If the temperatures drop during the night or high winds and strong rain occur, cover the plants to protect them during the harsh weather. Remove the protection when the weather improves and before the sunlight falls directly on the plants. Continue growing as you normally would through the growing season.

COMPOSITION AND MATERIAL OF TISSUE CULTURE

Tissue culture can be maintained either in liquid medium or semi-solid medium in the former, plant material is immersed in the medium either partially or completely while in the latter plant material is placed on the surface of the medium. The principal components of most plant tissue culture media are inorganic nutrients (macronutrients and micronutrients, carbon sources, organic supplements, growth relulators and a gelling agent

INORGANIC NUTRIENTS

A number of inorganic nutrients are required for normal growth of plants. Besides carbon, hydrogen, oxygen, other elements essential for plant growth include nitrogen, Phosphorucs, potassium, calcium, sulphur, magnesium, iron, manganese, copper, boron, zinc and molybdenum. Iron is added in the form of Ferric sulphate while nitrogen is furnished in the form of nitrate and ammonia. Of all the mineral nutrients, nitrogen is responsible for the most pronounced effects on growth and differentiation of cultured tissues.

Organic Nutrients

Cultured plants accomplish better growth when medium is supplemented with organic nutrients such as amino acids and vitamins. The most commonly used vitamin is thiamine (vitamin B_2). Other vitamins which improve the growth of cultured plants include nicotinic acid acid pantothenate and pyridoxine. Besides these boitin, folic acid and ammobenzoic acid are added in various concentrations in to some of the media. In order to promote growth of callus, some complex semi-synthetic substances are added in the medium. They include yeast extract (YE), coconut milk (CM), Casein Hydrolysate (CH), tomato juice (TJ) and malt extract (ME). As a carbon source, sucrose is the most preferred carbohydrate. Glucose, fructose, maltose, galactose, mannos and lactose are other favourable sugars.

Growth Hormones

Growth hormones helps in better plant development. The most commonly used auxins are IAA, IBA, NAA (naphthalene acetic acid) and 2,4-D (dichlorophenoxyacetic acid). These auxins promote cell division and root differentiation. Cytokinins are responsible for cell division and shoot differentiation. BAP (benzyl-amino purine), and 2-isopentanyl adenine are most widely used cytokinins. Among gibberellines, GA is commonly used.

Agar

Agar (a polysaccharide obtained from sea weeds) is used to solidify the medium. Agar is commonly used at a concentration of 0.8-1 per cent (W/v). Use of higher concentration of agar makes the medium hard This prevents the diffusion of nutrients into the tissues.

pH

The pH of the medium is adjusted between 5 to 5.8 by adding (0.1 N) NaOH or HO. Usually a pH higher than six results in a fairly hard medium whereas a pH below five does not allow satisfactory solidification of the medium.

MEDIA PREPARATION

The simplest method of preparing culture media is to use commercially available powdered media. These media contain all the requisite nutrients. However, agar, sugar and other supplements are added in the finally prepared media.

For the preparation of medium, the following steps are followed:

- Appropriate quantities of agar-agar and sucrose are dissolved in distilled water.
- Required quantities of stock solutions, growth hormones and other supplements are added
- More distilled water is added to make the final volume of the medium.
- pH of the medium is adjusted between 5-5.8 by adding 0.1 N NaOH or 0.1 N HCl.
- The medium is poured in culture tubes, flasks or any other containers.
- Culture vessels are plugged with non-absorbent cotton wool wrapped in cheese clothes. This will allow free gaseous exchange but inhibits microbial contamination.
- Culture vessels are sterilised by autoclaving at 120°C (1.06 kg/cm2) for 15-20 minutes.
- Culture medium is allowed to cool at room temperature and used or stored at 4°C.

INOCULATION AND CULTURING

The process of inoculation consists of sterilisation and isolation of single cells with which the medium is inoculated. This is followed by careful culturing the inoculum.

STERILISATION

Depending upon the type of sterilisation and material to be sterilised. There are four means of sterilisation.

Dry Heat

Glass wares (flasks, tubes, filters and pipettes) metal instruments and other articles which do not get charred by high temperature are put in special containers or wrapped in thick brown paper or thick aluminium foil and placed in drying oven and sterilised for a period of not less than 4 hours at a temperature of 140-160°C and then taken to the transfer chamber.

Wet Heat

In this process an autoclave operated with water vapour under pressure (steam pressure of 151b/inch2) and at a temperature of 120°C applied for 20 minutes when the autoclave has reached proper temperature mid the residual air enclosed inside the autoclave has been displaced by steam. "Wet" heat sterilisation process is recommended for heat stable solutions. Rubber items or other partly heat labile articles must not be autoclaved.

Filtration

Some protein material, vitamins and growth substances are heat labile and cannot be autoclaved and, therefore, they must be sterilised by filtration using ultrafilter at room temperature, before being added to an agar medium. The solution of these substances are added to the autoclaved medium with a sterile pipette while the agar medium is cooling and still in the sol state. In using this procedure the filter pore size is given utmost importance. A pore-size of 0.22 mm has been recommended in Millipore Catalogue and pure chasing guide 1978/79. All the stopcocks of the filtration unit are opened before it is wrapped in thick brown paper and autoclaved. Media or its constituents can also be sterilised by passing them through a pyrex sintered -glass sterilising filter (porosity H5).

Sterilisation by Chemicals

Surgical blades, scalpel are not sterilised by dry heat, because the high temperature makes the cutting edges dull. These articles as well as spatulas and forceps are usually immersed in 80 per cent V/v, ethyl alcohol until required, and sterilised during use by frequent immersion in alcohol and flaming. Before the aseptic removal of plant organ or tissue explant it is necessary to surface sterilise the plant material. Surface sterilisation may be accomplished by using an appropriate surface sterilant. Generally aqueous solutions of calcium or sodium hypochlorite ($Ca(OCl)_2$ or NaOCl), which release chlorine as the active sterilant are used. Some other chemicals which have also been used are H_2O_2, bromine water, silver nitrate and mercuric chloride.

ISOLATION OF SINGLE CELLS

Isolation of cells are done in several methods which are explained here.

From Plant Organs

The most suitable material for isolation of single cells is the leaf tissue since a more or less homogeneous population of cells are there in the leaves. There are good for raising defined and controlled large-scale cell cultures. From such intact plant organs (as leaf tissue) single cell can be isolated using mechanical or enzymatic method.

Mechanical Method

The procedure involves mild maceration of 10 g leaves in 40 m1 of the grinding medium (20 u mol sucrose, 10 u mol $MgCl_2$, 20u mol tris-HCI buffer with pH 7.8) with a mortar and pestle. The homogenate is passed through two layers of tmuslin cloth and the cells thus released are washed by centrifugation at low speed using the same medium. Isolation of free parenchymatous cells can also be achieved on a large scale by the mechanical method.

Enzymatic Method

Isolation of single cells by the enzymatic method has been found convenient as it is possible to obtain high yields from preparations of spongy parenchyma with minimum damage or injury to the cells. This can be accomplished by providing osmotic protection to the cells, while the enzyme macerozyme degrades the middle lamella and cell wall of the parenchymatous tissue. Applying the enzymatic method to cereals has proven difficult since the mesophyll cells of these plants are apparently elongated with a number of interlock constrictions, thereby preventing their isolation.

From Cultured Tissues

The most widely applied approach is to obtain a single cell system from cultured tissues. Freshly cut pieces from surface-sterilised plant organs are simply placed on a nutrient medium (solidified) consisting of a suitable proportion of auxins and cytokinins to initiate cultures. Explants on such a medium exhibit callusing at the cut ends, which gradually extends to the entire surface of the tissue.

The callus is separated from an explant and transferred to a fresh medium of the same composition to enable it to build up a mass of tissue. Repeated subculture on an agar medium improves the friability of the callus, a pre-requisite for raising a fine cell suspension in a liquid medium. The pieces of undifferentiated and friable callus are transferred in a continuously agitated liquid medium dispensed in autoclaved flasks or other suitable vials. Agitation is done by placing the culture flasks/vials on an orbital platform shaker or any other suitable device.

Movement of the culture medium exerts mild pressure on small pieces of tissue, breaking them into free cells and small cell aggregates. Further, it augments the gaseous exchange between the culture medium and the culture air, and also ensures uniform distribution of cells as well as cell clumps in the medium.

Growth and Maintenance

Cell suspensions are clonally maintained by the routine transfer (sub-culture) of cells in the early stationary phase to a fresh medium. During the incubation period the biomass of the suspension cultures increases due to cell division and cell enlargement This continues for a limited period since the viability of cells in suspension after the stationary phase de- creases due to the exhaustion of some factors or the accumulation of toxic substances in the medium. At this stage an aliquot of the cell suspension with uniformly dispersed free cells and cell aggregates is transferred to a fresh liquid medium of the: original composition. The timing of subcultures is very important.

The incubation period from cultures initiation to the stationary phase is determined primarily by:

- Initial cell density,
- Duration of lag phase and
- Growth rate of cell line.

The cell density used to subculture is critical and depends largely on the type of suspension culture to be maintained. Low initial cell densities will prolong the lag phase and exponential phases of growth. While initiating new suspension culture it is necessary to determine optimal cell density, proportionate to the volume of the culture medium, in order to achieve maximum growth.

At an initial cell density of 9-15 × 10^3 mI^{-1}, the cells will generally undergo an eightfold increase in cell number before entering the stationary phase. Subcultures established with a high inoculums rate (0.5 -2.5 × 10^5 cells ml^{-1}) show an increase in cell number during the incubation period to a range 1.4 × 106 mI^{-1} before entering the stationary phase.

The normal incubation time of stock cultures is 21-28 days between subcultures although cloning may occur within 18-25 days. In cases in which the cells are in a very active state of division, the passage length may be reduced to 6-9 days. Cell cultures initiated at very low cell densities will not grow unless the medium is enriched with the metabolites necessary to grow single cells or a small population of cells.

METHODS OF SINGLE CELL ISOLATION

From Plant Organ

The most suitable material for the isolation of single cells is the leaf tissue since a more or less homogeneous population of cells in the leaves offer good candidates for raising defined and controlled large scale cell culture. From such intact plant organs single cells can be isolated using mechanical or enzymatic methods.

Mechanical Method

Mechanical isolation involves tearing or chopping surface sterilised explant to expose the cells followed by scrapping of the cells with a fine scalpel to liberate the single cells hoping that it remained undamaged. Gnanam and Kulandaivelu (1969) developed a procedure to isolate mesophyll cells, active in photosynthesis and respiration, from mature leaves of several species of dicot and monocot. The procedure involves mild maceration of 10g leaves in 40 ml of grinding medium (20 μ mol sucrose, 10μ mol Mgcl2, 20 μ mol tris – HCL buffer, Ph7.8) with a mortar and pestle. The homogenate is passed through two layer of muslin cloth and cell thus released are washed by centrifugation at low speed using the same medium. The mechanical isolation of free parenchymatous cells can also be achieved on a large scale.

Enzyme Method

Takebe et al. (1968) treated tobacco leaf tissue with enzyme pectinase and obtained a large number of metabolically active cells. The potassium dextran sulphate in the enzyme mixture improved the yield of free cells.

Isolation of single cells by the enzymatic method has been found convenient, as it is possible to obtain high yields from preparation of spongy parenchyma with minimum damage or injury to the cells. This can be accomplished by providing osmotic protection to the cells while the enzyme macerozyme degrades the middle lamella and cell wall of the pranchymatous tissue. Applying enzymatic method to cereals (Hordeum vulgare, Zea mays) has proven rather difficult since the mesophyll cells of these plants are apparently elongated with number of interlocking constrictions, thereby preventing their isolation.

From Cultured Tissue

The most widely applied approach is to obtain a single cell system from cultured tissue. Freshly cut piece from surface sterilised plant organs are simply placed on a nutrient medium consisting of a suitable proportion of auxins and cytokinins to initiate cultures. Explant on such a medium exhibit callusing at the cut ends, which gradually extends to the entire surface of the tissue.

The callus is separated from an explant and transferred to a fresh medium of the same composition to enable it to built up a mass tissue. Repeated sub-culture on an agar medium improves the friability of the callus, a pre-requisite for raising a fine cell suspension in a liquid medium. The piece of undifferentiated and friable callus are transferred in a continuously agitated liquid medium consisting of flask or suitable vials. Agitation is done by placing the culture medium exerts mild pressure on small pieces of tissue, breaking them into free cells and small cell aggregates. Further, it augments the gases exchange between the culture medium and the culture air and also ensures uniform distribution of cells as wells as clumps in the medium.

METHODS OF TISSUE CULTURE STUDY

Process of biotechnology in tissue culture may be broadly placed in four groups.

Somatic Cell Fusion

This is achieved by inducing two or more pro top lasts to fuse and the fusion product is nurtured to produce a hybrid plant. Though, this phenomenon is of common occurrence, protoplasts of some plant species fail to fuse.

Some scientists have shown the production of hybrids that cannot be produced by the conventional hybridisation methods because of-sexual or physical incompatibility among certain plants.

Genetic Engineering

Experiments are being conducted by some scientists to introduce nuclei, chloroplasts, viruses, DNA, mitochondria, plastids, etc., keeping in view the capability of isolated protoplast to ingest "foreign" bodies by a process similar to the process of endocytosis often seen in cells of protozoan and other animals.

Wall Biosynthesis

Scientist have made use of isolated protoplast for the study of wall biosynthesis and deposition because of the ability of cultured protoplasts to regenerate a cell wall rapidly.

Study of Protoplast Population

As the protoplasts can be manipulated in a manner similar to that of the micro-organism, selection of mutant cell lines and the cloning of cell population is possible by using microbiological methods.

FUNCTION OF TISSUE CULTURE

In the recent years plant protoplast, cell and tissue cultures have become an important tool for crop improvement, commercial production of natural compounds and in the development of forestry.

Some of the areas of biotechnological application of cultured plant protoplasts/ cells/tissues are:

- Tissue culture applications in order to capitalise on the totipotency of cells,
- Cells and protoplast cultures coupled 'with DNA vectors to overcome problems caused by barriers to gene transfer through sexual means,
- Culture of plant cells for the production of useful compounds,
- Extension and increase of deficiency of biological nitrogen fixation, and
- Transfer of genes for nitrogen fixation ability to non-fixing species.

Due to the manipulation of plant tissues in the laboratory this technique has been referred to by some researchers, as a 'botanical laser' whose numerous uses are yet to be fully understood.

APPLICATION IN AGRICULTURE

The major areas of application of tissue culture are briefly highlighted here.

Improvement of Hybrids

Development of cell fusion and hybridisation techniques have solved the problem of incompatibility of plants and widened the scope of production of new varieties with in a short time. For example breeders obtained somatic hybrid of wild and cultivated potatoes (*S. tuberosum, S. chacoense*) and succeeded in the induction or organogenesis.

The somatic hybrid plant inherited many characters, *viz.*, intermediate leaf morphology, stomata, forms and colour of tubers prolonged flowering, large and fertile pollen grains, high yield, resistance against Y-virus. However, haploid plant materials available as protoplast, cell and tissue culture systems are currently being evaluated for the use in transfer of foreign genetic material to selected plant species by protoplast fusion, transformation, transduction and organelle transfer.

Production of Encapsulated Seeds

The two 'techniques, somatic embryogenesis and organogenesis are the alternative methods for regeneration of the whole plant from cultured tissue *in vitro*. Moreover, the recent works on the wrapping an *in vitro* derived embryos in a seed coat, *i.e.*, production of synthetic seeds are important because it may improve agriculture due to its low cost.

Production of Disease Resistant Plants

Many plant species, which propagate vegetatively are systematically infected by virus, bacteria, fungi and nematodes. Their inoculum is carried over several generations resulting in continued adverse effect of productivity and quality of crops. In order to ensure highest possible yield and quality, it is necessary to provide disease free stock plants to growers. Tissue culture techniques have solved this problem and minimized the time of biological testing.

Unless large scale population of pure inoculum of test pathogens are available, it is difficult to pursue the establishment of pathogenicity and crop loss-assessment as it is done in field conditions. Now it has become possible to carry out such experiments in laboratory within short span of time by using tissue culture technologies. Scientists have discussed the following advantages for the study of several aspects of host-pathogen interactions and responses.

They are:

- Ability to isolate host cells without wounding,
- Control of inoculum of pathogen and number of host cells,
- Ability t'J change the nature of host pathogen interaction by altering the constituent of growth medium,

- Presence of only one or a few major host cell types and
- Easy to apply and remove materials, *e.g.*, labelled precursors form cultured cells.

Production of Stress Resistance Plant

Biochemical mechanisms exist is Culture cells which determine the resistance to biocide chemicals and provide the theoretical promise for selection *in vitro*. For example: cell suspension tolerant to 2,4-D when the cells were subcultured for six months in liquid medium supplemented with increasing amount of herbicides. The cells were able to grow in one rnM (milli mol) 2,4-D, while the control suspension was completely inhibited at O.3mM 2,4-D. It is suggested that tolerance was the result of induction of enzymatic systems responsible for the degradation of 2,4-D.

Uses of protoplasts as a system to select cell lines tolerant to herbicides have not been extensively explored. Protoplast technology can be used to increase the possibility to obtain monoclonal lines and offer the opportunity for intraspecific\transfer of cytoplasmic factor of resistance to some type of herbicide. A potential application of transfer of herbicides tolerance factor is the transplant of cytoplasmic organelles, for l example chloroplasts.

Transfer of Nif Gene to Eukaryotes

Nitrogen fixing ability, a genetic character, exists in prokaryotic diazotrophs. However, one of the major tasks is the transfer of this character to eukaryotes. In recent years, researches are being done to solve this problem through tissue culture techniques coupled with the recombinants DNA technology.

Historically, nitrogen fixation by rhizobia was believed to occur through symbiosis. For the first time an excitement was caused in scientific community with the discovery by Holsten *et al.* (1971). They obtained active rhizobia in the absence of nodules, leghaemoglobin and bacteroids which were apparently necessary in the intact plants.

They established *Rhizobium joponicum.* In cell suspension of soybean roots. Callus induced from root explants of soybean on a specific medium was inoculated with R. joponicum. Later on it was microscopically observed that infection threads were formed by the bacteria which were present between intracellular spaces.

They multiplied inside cells. Moreover, development of nitrogenase in soybean callus- Rhizobium system growing on solid was also observed and it was also found that only specific (isomorphic) cells of callus are vulnerable to infection by the bacteria.

In addition to improvement the bacterial strains and increased nodulation, it is necessary to seek those genotypes with the efficient photosynthesis and improved partitioning of carbohydrates to nodules.

Future Prospects

In addition to work done successfully on nif gene transfer, there are other important genes which have been clones, for example:

- Phaseolin and leghaemoglobin genes in soyabean;
- Storage protein genes in soybean,
- Genes of ribulose bi-phosphate carboxylase/oxygenase (Ru BP case) of pea, maize, wheat, etc.

Success achieved on these aspects would certainly promote in green revolution. Moreover, improvement in primary productivity by conversion of C_3 plants into C_4 ones through genetic engineering techniques hopefully would increase the primary productivity.

APPLICATION IN HORTICULTURE AND FORESTRY

Tissue culture is applied in a number of cases in horticulture and forestry.

Micropropagation

Microprogation is a perfect alternative to asexual propagation of important plants and trees, because of following reasons:

- In this method only a small piece of tissue is needed to generate millions of clonal plants a year
- This method provides a possible alternative method for developing resistance in many species.
- It solved problem of quarantine for introduction of art disease in new areas and provide a mean for international exchange of plant material.
- By this method multiplication can be made in any season.

Regeneration of plantlets from cultured plant cells and tissues has been achieved in many trees of high economic value. Many of the studies are aimed at large scale micropropagation of important trees yielding fuel, pulp, timber, oils or fruits. Therefore, clonal forestry and horticultural are gaining an increasing recognition as an alternative for tree improvement. However, strategies for transferring cultured plants *in vitro* to field conditions are based on relatively higher priced horticultural species rather than agricultural and forestry species.

At planting out stage the plantlets fail to survive because of sudden change in the environment and invasion by soil microbes. So the regenerates should be transferred first to green house and then to field. The humidity should be controlled by covering the plants with transparent polythylene sheets. Some important plants on which work is done are:

Acacia nilotica, Albizia lebback, Aprocera, Azadirachta Indica, Bauhinia purpurea, Butea monosperma, Dalbergia spp, Dendrocalmus strictus, Eucalyptus spp, Ficus religiosa, Morus spp, Populus spp, Shorea robusta, Tectona grandis, Cryptomeria japonica, Picea smithiana and Pinus spp (all gymnosperms).

In Vitro Establishment of Mycon-biza

Mycorrhizal fungi show highest specialisation of parasitism. But major problems with them is their failure to grow on an artificial medium in laboratory. Therefore, establishment and multiplication of mycorrhizal fungi on cultured tissue of the same host plant, if successfully developed, may be a good tool for handling mycorrhizal fungi, production of high potential inoculum and their establishment in root systems of nursery plants in horticulture and forestry and the plantation of mycorrhiza infested seedlings into the field.

Many attempts have been made to establish vesicul arbuscular mycorrhizal (VAM) fungi in axenic culture but unfortunately none of them got success. Moreover, micorhizal fungi have been cultured only on cortical tissues of roots which were seperated from the whole plant, as in root organ cultures where it acted as food base. VAM fungi have very high degree of specialisation for food base on root cortex.

APPLICATION IN INDUSTRIES

Production of useful compounds by cultured plant cells has become a field of special interest in various biotechnological programmes. However, much attention has been paid on the production of pharmaceutical and other secondary compounds such as essential oils, food flavourings and colourings which are used- by the fast food, ice cream and confectionary industries. This can be achieved by selection of specific cells producing high amount of desired compounds and the development of a suitable medium. In general, secondary metabolites produced by plant cell cultures are rather in small amount but strains of cells producing the same are in greater amount than those found in the intact plants have been isolated by clonal selection.

Commonly two methods are employed for the selection of specific cells: Single cell cloning and cell aggregate cloning. The difficulties associated with isolation and culture of single cells limit application of this method. The later may appear to be more time-taking but easier than the first one. Recently, *in vitro* production of high amount of useful compounds has increased with the success obtained so far in experimental studies. It is hoped that in near future, the industrial production of such compounds by using these techniques would be possible. Similarly, monogenic cultures of nematodes have been used for the study of mechanisms of action of nematicide. This technique can be used extensively in industry to supply nematodes for nematicide screaning programmes.

ORGANISMS OF PLANT CELL ORGANELLES

Plants are highly evolved eukaryotic organisms that comprise of membrane bound cell organelles. Even though plants and animals belong to eukaryotic groups, they differ in certain characteristic features. For example, a plant cell possesses a well-developed cell wall and large vacuoles, while an animal cell lacks such structural parts. In addition to these, a plant cell lacks centrioles and intermediate filaments, which are present in an animal cell.

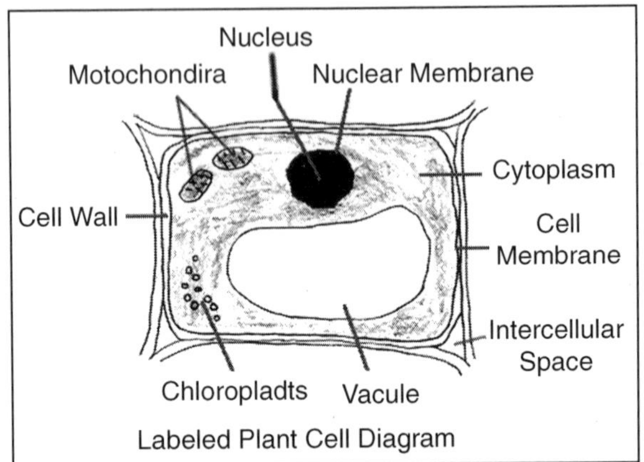

Labeled Plant Cell Diagram

A typical plant cell is made up of cytoplasm and organelles. Scientific studies have been done regarding plant cell organelles and their functions. Each of the organelles of a plant cell has specific functions, without which the cell cannot operate properly.

LIST OF PLANT CELL ORGANELLES

When it comes to plant cell organelles, they are more or less similar to animal cells, except that the latter lacks chloroplast organelles, that are responsible for photosynthesis.

Following is a list of organelles found in plant cell:

Nucleus

Nucleus (plural nuclei) is a highly specialized cell organelle, which stores the genetic component (chromosomes) of the particular cell. It serves as the main administrative centre of the cell, by coordinating the metabolic processes like cell growth, cell division and protein synthesis.

Plastids

Plastids are collective terms for organelles that carry pigments. In a plant cell, chloroplasts are the most prominent forms of plastids, that contain the green chlorophyll pigment. Because of these chloroplast plastids, a plant cell has the ability to undergo photosynthesis in the presence of sunlight and synthesize its own food.

Ribosomes

Ribosomes are plant cell organelles that comprise of proteins (40 per cent) and ribonucleic acid or RNA (60 per cent). They are important organelles responsible for the synthesis of proteins. Each ribosome consists of two parts, a larger subunit and a smaller subunit.

Mitochondria

Mitochondria (singular mitochondrion) are spherical to rod-shaped organelles present in the cytoplasm of the plant cell. They break down the complex carbohydrates and sugars into usable forms, for the plant. As mitochondria indirectly supply energy for the plant cell, they are also called as the powerhouse of the cell.

Golgi Body

Golgi body is also referred to as golgi complex or golgi apparatus. It plays a major role in transporting chemical substances in and out of the cell. After the endoplasmic reticulum synthesizes lipids and proteins, golgi body alters and prepares them for exporting outside the cell.

Endoplasmic Reticulum

Endoplasmic reticulum (ER) is the connecting link between the nucleus and cytoplasm of the plant cell. Basically, it is a network of interconnected and convoluted sacs that are located in the cytoplasm. Based on the presence or absence of ribosomes, ER can be of smooth or rough types.

The former type lacks ribosomes, while the latter is covered with ribosomes. Overall, endoplasmic reticulum serves as a manufacturing, storing and transporting structure for glycogen, proteins, steroids and other compounds.

Vacuoles

Vacuoles are the storage organelles that help in regulating turgor pressure of the plant cell. In a plant cell, there can be more than one vacuole. However, the centrally located vacuole is larger than others, which stores all sorts of chemical compounds. Vacuoles also assist in intracellular digestion of complex molecules and excretion of waste products.

Peroxisomes

Peroxisomes are cytoplasmic organelles of the plant cell, which contains certain oxidative enzymes. These enzymes are used for the metabolic breakdown of fatty acids into simple sugar forms. Another important function of peroxisomes is to help chloroplasts in undergoing photorespiration process.

Well! This was a brief information regarding plant cell organelles, their structure and specific functions. As we have seen, coordination of the functions of plant cell organelles is crucial for carrying out the physiological and biochemical functionalities of the plant.

Cell Nucleus: Structure

The structure of a cell nucleus consists of nuclear membrane (nuclear envelope), nucleoplasm, nucleolus and chromosomes. Nucleoplasm, also known as karyoplasm, is the matrix present inside the nucleus. Let's discuss in brief about the several parts of a cell nucleus.

Nuclear Membrane

The nuclear membrane is a double-layered structure that encloses the contents of the nucleus. The outer layer of the nuclear membrane is connected to the endoplasmic reticulum. A fluid-filled space or perinuclear space is present between the two layers of a nuclear membrane. The nucleus communicates with the remaining of the cell or cytoplasm through several openings called nuclear pores. Nuclear pores are the sites for the exchange of large molecules (proteins and RNA) between the nucleus and cytoplasm.

Chromosomes

Chromosomes are present in the form of strings of DNA and histones (protein molecules) called chromatin. Chromatin is further classified into heterochromatin and euchromatin based on the function. The former type is a highly condensed trancriptionally inactive form, mostly present in adjacent to the nuclear membrane. Euchromatin is a delicate, less condensed organization of chromatin, which is found abundantly in a transcribing cell.

Nucleolus

The nucleolus is a dense, spherical-shaped structure present inside the nucleus. Some of the eukaryotic organisms have nucleus that contains up to four nucleoli. The nucleolus plays an indirect role in protein synthesis by producing ribosomes. Ribosomes are cell organelles made up of RNA and proteins; they are transported to the cytoplasm, which are then attached to the endoplasmic reticulum. Ribosomes are the protein-producing structures of a cell. Nucleolus disappears when a cell undergoes division and is reformed after the completion of cell-division.

Cell Nucleus: Functions

Speaking about the functions of a cell nucleus, it controls the hereditary characteristics of an organism and is responsible for the protein synthesis, cell division, growth and differentiation.

Here is a list of the functions carried out by a cell nucleus:

- Storage of hereditary material, the genes in the form of long and thin DNA (deoxyribonucleic acid) strands, referred to as chromatins.
- Storage of proteins and RNA (ribonucleic acid) in the nucleolus.
- Nucleus is a site for transcription in which messenger RNA (MRNA) are produced for the protein synthesis.
- Exchange of hereditary molecules (DNA and RNA) between the nucleus and rest of the cell.
- During the cell division, chromatins are arranged into chromosomes.

- Production of ribosomes (protein factories) in the nucleolus.
- Selective transportation of regulatory factors and energy molecules through nuclear pores.

As the nucleus regulates the integrity of genes and gene expression, it is also referred to as the control centre of a cell. Overall, the cell nucleus stores all the chromosomal DNA of an organism.

The environment which we inhabit and live our daily lives, is an incorporation of several kinds of components. These components, although may not share any resemblance, yet are in a mutual dependency with each other.

These components not only refer to the living things, but other elements which are classified as the non-living things on the Earth. Both these elements are important in their own domain in the environment, as they work to maintain the delicate balance in the ecosystem.

And this very balance helps to regulate all those things which makes life possible on the Earth. Ergo, with any of these things missing in this system, the biological organization will suffer nothing less than chaotic events. So let's proceed further to understand these types of components and what role do they play in the levels of organization of life

LEVELS OF ORGANIZATION

In an ascending manner, the levels of biological organization are as follows:

Subatomic Particles

Any substance is composed mainly of three subatomic particles. They include the protons, neutrons and electrons. As you must be aware, protons refer to positively charged particles, while neutrons are ones which posses no charge. These two kinds reside inside the nucleus of the atom. Now, about the electrons, they posses negative charge and rotate around the nucleus, in different energy shells. Another subatomic particle is what is known as the photon - possessing zero mass and rest energy and traverses with the speed of light. It is actually defined as a quantum of electromagnetic energy.

Atoms

After the subatomic particles, now we come to the next point in the levels of organization, know as atoms. All objects are made up of matter in this universe. And the basic building blocks of these matter are known as atoms. These components consist of equal number of protons and neutrons. However, a single element might have two different atoms, wherein, there might be a difference in the number of neutrons. The centre of the atom is known as the nucleus, and it comprises of protons and neutrons. This structure of the atomic nucleus is orbited by negatively charged electrons.

Small Molecules

By saying small molecules, I am referring to those levels of organization, which comprises of amino acids, fatty acids, glucose, etc.

Macromolecules

Now these are the large molecules which are contained in a cell, and are required for carrying out the vital processes of life. Examples include carbohydrates, lipids, proteins, nucleic acids, etc.

Molecular Assemblies

A level higher than macromolecules, these assemblies are defined as sets which comprise of one or more molecular entity.

Organelle

Every cell has a specialized part which serve as organs. These are known as organelle. Examples are nucleus, mitochondria, etc.

Cells

This is one of the important levels of organization, wherein, cells are known to play vital roles. Every living creature on this Earth consists of cells, which are primarily, their structural and functional units. For instance, microorganisms like bacteria are known to be single-celled organisms. While, we humans and other organisms, are known to have as many as 100,000,000,000,000 cells!

Tissues

Tissue is nothing but a term that is used to refer to a group of cells. There are several functions in the body of living organisms. And different function is assigned to different tissues. For example, muscle tissue, nervous tissue, etc., are found in animals, while plants, have meristematic tissues and permanent tissues.

Organs

Again, when tissues are grouped together, they form what is known as an organ, which also has an important place in the levels of organization. As you are aware, heart, lungs, kidneys, stomach are organs. And these organs are enabled by the different tissues to carry out their specific duties.

Organ System

As you must have guessed already, this system is a formed by different kinds of organs. The system is assigned to perform higher levels of work in the body. This is more important in multicellular organisms like us and animals. A very simple example of an organ system is the circulatory system. It is a collaboration of heart, blood and blood vessels.

Organisms

Now when all the above levels of organization are taken in one picture, what is derived is known as organism like plants, animals, humans, bacterium, etc.

Population

It is here that we speak about a broader concept in the levels of organization of living things. A group of organisms, belonging to the same species may have the capability to inter-breed with each other. And when they inhabit a particular area, they are known as population. Its example can be the population of frogs in a pond.

Species

Species is a term that is used to mark the distinction between different kinds of organisms. For example, human species are different from animal species and species of plants.

Community

Taking it down to the last three of the levels of organization, we are speaking of community. It is nothing but a reference to a group of organism of different species, which populate a given area and stay in mutual interaction with each other.

Ecosystem

To define it in a layman language, interaction of a group of organism with the environment that they live in, is an ecosystem. For example, a pond ecosystem may consists of organisms such as fish, frogs, insects, etc., who depend on the water, mud, plants, etc., for survival.

Biosphere

Finally, we have reached the last block of the pyramid of the levels of organization, with what is known as biosphere. It encompasses all. Meaning, it consists of the regions which lie on, above and below the surface of the Earth, including the atmosphere where the living organisms thrive. So there you are with a quick explanation on the different levels of organization, which we, along with several other components are living. As we can see, each level serves as a building block for the other, and this is what lays emphasis on the importance of all the components.

TOTIPOTENCY OF CELLS

Totipotency is the ability of all living cells potentially to regenerate whole new individuals - the trick is how to trigger them to do it and so to demonstrate it. If I remember correctly, Dolly the sheep was cloned from a cell from another sheep's udder - a differentiated cell, which in normal circumstances would have remained as such. Similarly in plants, it is possible to trigger single epidermal cells in oilseed rape, sunflower, and I am sure there are others, to develop into embryos and subsequently to whole plants.

Even immature pollen grains have this potential - these immature cells can be switched to develop into whole plants, even though they only have one set of chromosomes, and give rise to haploid plants, although sometimes these may diploidise during culture, to generate homozygous diploid plants. One of the hardest things in plant tissue culture is to get tissues from mature trees to illustrate totipotency, but by forcing a rejuvenation step, it is even possible to do this with some species - I believe quite a bit of work has been carried out on pines.

Simple cuttings, and the SAPS tissue culture system depends on this ability of cells to switch their development - hence root formation from hypocotyls, and root formation from woody cuttings, which allow new plants to be obtained for propagation. We described the different media used for cell/tissue culture, and also discussed culture of individual differentiated cells. However, in plant tissue culture experiments, unless single cell cultures are absolutely necessary, more often we use explants (pieces of differentiated tissues) to initiate their growth in culture.

This is also necessary to obtain callus from which single cell cultures are derived. During the last few decades, even for development of new crops or for improvement in the characteristics of existing crops (e.g. to obtain disease free crop, etc.), the techniques of tissue culture in combination with genetic manipulation have become important tools. For instance, as an alternative to vegetative propagation, shoot tip propagation was used commercially for rapid and consistent reproduction of elite or difficult to propagate genotypes. , When an explant from differentiated tissue is used for culture on a nutrient medium, the non-dividing, quiescent cells first undergo certain changes to achieve a meristematic state.

The phenomenon of the reversion of mature cells to the meristematic state leading to the formation of callus is called dedifferentiation. The component cells of callus have the ability to form a whole plant, a phenomenon described as redifferentiation. These two phenomenons of dedifferentiation and redifferentiation are inherent in the capacity described as cellular totipotency, a property found only in plant cells and not in animal cells.In other words, while a differentiated plant cell retains its capacity to give rise to a whole plant, an animal cell loses this capacity of regeneration after differentiation.

Although, generally a callus phase is involved before the cells can undergo redifferentiation leading to regeneration of a whole plant, but rarely the dedifferentiated cells can give rise to whole plants directly without an intermediate callus phase. These aspects of tissue culture and its application in clonal propagation and in the general on of hereditary variation called somaclonal variation.

Some plants have cells that are able to convert from a structured, or differentiated, state to a totipotent state. One example is the ZZ plant (Zamioculcas zamiifolia). This plant is so cellularly precocious that a mere piece of leaf or stem can be used to grow a new plant. Other plants and trees can produce spontaneous buds on the stem or trunk. These emerge from the activity of the meristematic tissue of the cambium layer, and since they are not buds that are normally produced in the axils of leaves, they are called adventitious buds. This can happen even when the tree in question is not the kind that will root from cuttings. The formation of adventitious buds begins with the production of a callus, which is essentially an undifferentiated lump of cells resembling a small tumor. Through genetic and hormonal processes, cells differentiate into growing points which then produce new stems with leaves. A similar process can result in adventitious roots as well.

Advances in the understanding of totipotency and how cells can be stimulated to become totipotent has resulted in an entire industry, the business of micropropagation, or what is commonly referred to as "tissue culture". While micropropagation labs do culture tissue (meristematic tissue), this work is not the same as what is done in a scientific tissue culture laboratory. The work done in scientific labs is much more esoteric and amazing. I studied the science of tissue culture when I attended graduate school, and we did things like isolating single cells from a callus and removing the cell wall using enzymes to yield a naked plant cell called a protoplast.

These are important in scientific research because genes are more easily inserted into plant cells when they don't have a cell wall around them. We grew pure callus cultures and cell lines that were, essentially, all totipotent cells. Using the right combinations of plant hormones, we could then cause the cells to turn into plant shoots with roots. These and other experiments done in scientific tissue culture labs are not the kinds of activities engaged in by micropropagation operations, but they do demonstrate vividly the power of totipotency in plant growth.

By contrast, commercial tissue culture is the business of producing many thousands of clones of particular plants using meristematic tissue. Different plants require different formulations of media, as the formula that is ideal for one plant will produce no results with some other plants. Conventional micropropagation requires a sterile or aseptic environment and the operating costs of such a facility are substantial. However, without totipotency, none of this would be possible at any cost.

FORMATION OF THE EMBRYO

A plant reproduces naturally through the development of zygotic embryos. Formation of the embryo begins with the division of the fertilized eggs or zygote within the embryo sac of the ovule. Through an orderly progression of divisions, the embryo eventually differentiates, matures , and develops into the new plantlet. Alternatively, the plant can be derived from a single somatic cell or a group of somatic cells. This regeneration process, which differs from the natural pathway, is called somatic embryogenesis. All the plantlets produced have the same genetic makeup. Combined with genetic engineering, micropropagation through somatic embryogenesis provides an efficient means of producing a large number of elite or transgenic plants.

The somatic embryo's development is analogous to its zygotic counterpart. The embryo development process is characterized by a series of morphological changes which correspond to its maturition. Research in our lab focuses on the applications of image analysis and pattern recognition systems for studies of somatic embryo development. Carrot embryos have been used as a model system to develop the image analysis and pattern recognition system. Currently, Douglas-fir cultures are being investigated because of their economic importance to the timber industry.

6

Technology of Gene Transfer in Plant Science

METHODS OF GENE TRANSFER IN PLANTS

To add a desired trait to a crop, a foreign gene (transgene) encoding the trait must be inserted into plant cells, along with a "cassette" of additional genetic material. The cassette includes a DNA sequence called a "promoter," which determines where and when the foreign gene is expressed in the host, and a "marker gene" that allows breeders to determine which plants contain the inserted gene by screening or selection. For example, marker genes may render plants resistant to antibiotics that are not used medically (*e.g.*, agromycin, canamycin) or tolerant to certain herbicides.

Two methods are used to transfer foreign genes into plants. The first method involves the use of a plant pathogen called Agrobacterium tumefaciens, which causes crown gall disease in many species. This bacterium has a plasmid, or loop of non-chromosomal DNA, that contains tumour-inducing genes (T-DNA), along with additional genes that help the T-DNA integrate into the host genome. For genetic engineering purposes, Agrobacterium must first be "disarmed" so that it does not make the plant sick. This is done by removing most of the T-DNA while leaving the left and right border sequences, which integrate a foreign gene into the genome of cultured plant cells.

The second delivery method is a "gene gun," which fires gold particles carrying the foreign DNA into plant cells. Some of these particles pass through the plant cell wall and enter the cell nucleus, where the transgene integrates itself into the plant chromosome. Because both methods of gene transfer are fairly random, one must screen for the plant cells that contain the foreign gene.

PLANT TRANSFORMATION METHODS

Plant transformation was first described in tobacco in 1984. Since that time, rapid developments in transformation technology have resulted in the genetic modification of many plant species. Methods for introducing diverse genes into plant cells include *Agrobacteriumtumefaciens*-mediated transformation, recently reclassified as *Rhizobium radiobacter,* direct gene transfer into protoplasts, and particle bombardment.

Gene Transformation

Several gene transformation techniques utilise DNA uptake into isolated protoplasts mediated by chemical procedures, electroporation, or the use of high-velocity particles (particle bombardment). Direct DNA uptake is useful for both stable transformation and transient gene expression. However, the frequency of stable transformation is low, and it takes a long time to regenerate whole transgenic plants.

Chemical Procedures

Plant protoplasts treated with polyethylene glycol more readily take up DNA from their surrounding medium, and this DNA can be stably integrated into the plant's chromosomal DNA. Protoplasts are then cultured under conditions that allowed them to grow cell walls, start dividing to form a callus, develop shoots and roots, and regenerate whole plants.

Electroporation

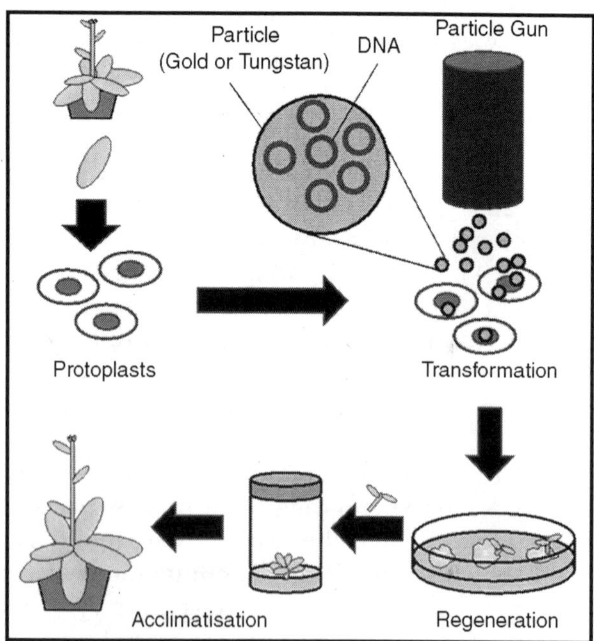

Fig. Plant Transformation Process Using Particle Bombardment Includes the following Steps: (1) Isolate Protoplasts from Leaf Tissues. (2) Inject DNA-coated Particles into the Protoplasts Using Particle Gun. (3) Regenerate into Whole Plants. (4) Acclimate the Transgenic Plants in a Greenhouse.

Plant cell electroporation generally utilises the protoplast because thick plant cell walls restrict macromolecule movement (Bates, 1999). Electrical pulses are applied to a suspension of protoplasts with DNA placed between electrodes in an electroporation cuvette. Short high-voltage electrical pulses induce the formation of transient micropores in cell membranes allowing DNA to enter the cell and then the nucleus.

Particle (Microprojectile) Bombardment

Particle bombardment is a technique used to introduce foreign DNA into plant cells. Gold or tungsten particles (1–2 μm) are coated with the DNA to be used for transformation. The coated particles are loaded into a particle gun and accelerated to high speed either by the electrostatic energy released from a droplet of water exposed to high voltage or using pressurised helium gas; the target could be plant cell suspensions, callus cultures, or tissues. The projectiles penetrate the plant cell walls and membranes. As the microprojectiles enter the cells, transgenes are released from the particle surface for subsequent incorporation into the plant's chromosomal DNA.

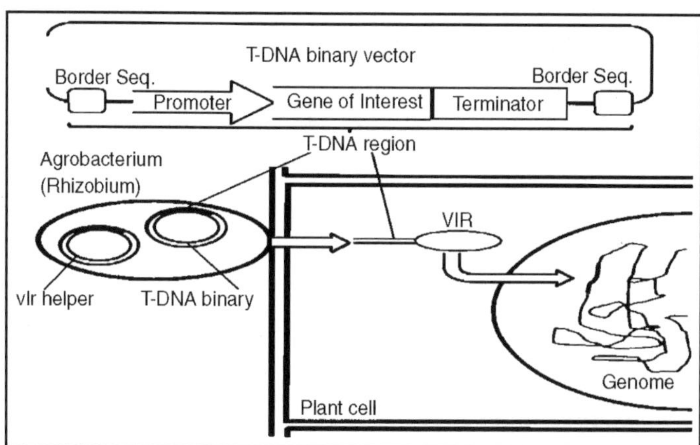

Fig. The *Agrobacterium*-mediated Transformation Process Includes the Following Steps: (1) Isolate Genes of Interest from the Source Organism. (2) Insert the Transgene into the Ti-plasmid. (3) Introduce the T-DNA Containing-plasmid into *Agrobacterium*. (4) Attach the Bacterium to the Host Cell. (5) Excise the T-strand from the T-DNA Region. (6) Transfer and Integrate T-DNA into the Plant Genome.

Using Agrobacterium for Plant Transformation

Agrobacterium-mediated transformation is the most commonly used method for plant genetic engineering. The pathogenic soil bacteria *Agrobacterium tumefaciens* that causes crown gall disease has the ability to introduce part of its plasmid DNA (called transfer DNA or T-DNA) into the nuclear genome of infected plant cells.

Transforming Arabidopsis Thaliana

Arabidopsis thaliana, a small flowering plant, is a model organism widely used in plant molecular biology. The first *in planta* transformation of *Arabidopsis* included the use of tissue culture and plant regeneration (Feldmann& Marks, 1987). The *Agrobacterium* vacuum (Bechtold *et al.*, 1993) and floral dipping (Clough & Bent, 1998) are efficient methods to generate transgenic plants. They allow for plant transformation without the need for tissue culture.

The floral dipping method markedly advanced the ease of creating *Arabidopsis* transformants, and it is the most widely used transformation method. These methods were later simplified and substantially improved, significantly reduced the required labour, cost, and time, as compared with earlier procedures. However, these transformation methods have some problems. The floral dipping method involves dipping *Arabidopsis* flower buds into an *Agrobacterium* cell suspension, requiring large volumes of bacterial culture grown in liquid media.

The large shakers and centrifuges, necessary to house the media, require sufficient experimental space. These factors limit transformation quantities. Here, we describe an improved method for *Agrobacterium*-mediated transformation that does not require the large volumes of liquid culture necessary for floral dipping.

Improved Technique for Agrobacterium-Mediated Transformation

A. thaliana can be stably transformed with high efficiency using T-DNA transfer by *Agrobacteriumtumefaciens*. *Agrobacterium*-mediated transformation using the floral dipping method is the most widely used method for transforming *Arabidopsis*. We have showed that *A. thaliana* can be transformed by inoculating flower buds with 5 μl of *Agrobacterium* cell suspension, thus avoiding the use of large volumes of *Agrobacterium* culture. Using this floral inoculating method, we obtained 15–50 transgenic plants per three transformed *A. thaliana* plants.

The floral inoculating method can be satisfactorily used in subsequent analyses. This simplified method, without floral dipping, offers an equally efficient transformation as previously reported methods. This method reduces overall labour, cost, time, and space. Another important aspect of this modified method is that it allows many independent transformations to be performed at once.

Agrobacterium Strains

The *Agrobacterium* strain GV3101 (C58 derivative) is frequently used to transform many binary vectors, *e.g.*, pBI121, pGPTV, pCB301, pCAMBIA, and pGreen, into *Arabidopsis*. It carries rifampicin resistance (10 mg l^{-1}) on the chromosome (Koncz & Schell, 1986). On the other hand, LBA4404 is a popular strain for tobacco transformation but is less effective for *Arabidopsis*.

Simplified Arabidopsis Transformation: Floral Inoculating Method

Until now, a limited number of constructs could be transformed into *Arabidopsis* because of difficulty growing large volumes of *Agrobacterium*. Therefore, we focused on improvements to the floral dipping method. The problem of space and volume can be solved by using a small culture volume. Each plant is transformed using only 30–50 μl of bacteria grown in 2 ml of liquid culture. Our present method, as described below, is a simple modification of the method reported by Clough and Bent (1998).

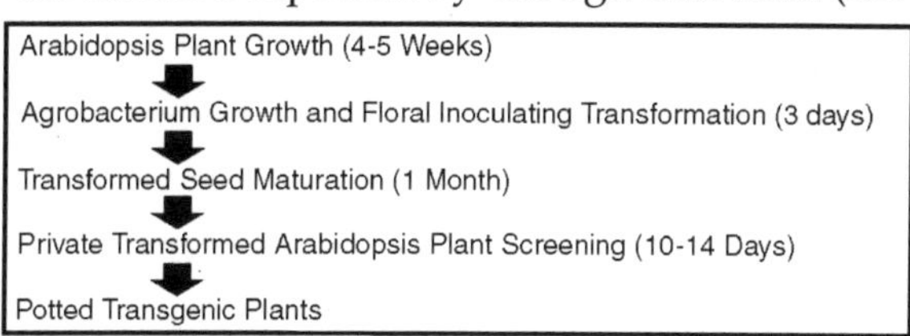

Fig. Transformation Using *Agrobacterium* and the Floral Inoculating Method.

Recent papers (Liu *et al.*, 2008; Zhang *et al.*, 2006) illustrate the floral dipping process. Clough and Bent (1998) reported that neither Murashige and Skoog (MS) salts and hormones nor optical density (OD) makes a difference in transformation efficiency. An *Agrobacterium* cell suspension containing 0.01–0.05 per cent Silwet L-77 (vol/vol) was used in the uptake of *Agrobacterium* into female gametes, instead of vacuum-aided infiltration of inflorescences.

Reagents

- *A. thaliana*: There are marked differences in transformation efficiency between various ecotypes. For floral dipping transformation, efficiency in the *Landsbergerecta* (*Ler*-0) ecotype is lower than that in the Columbia (Col-0) ecotype. Transformation efficiency in Wassilewskija (Ws-0) is very high among *Arabidopsis* ecotypes.
- *Agrobacterium strain:* GV3101 (Koncz& Schell, 1986) (recommended) or others.
- 0.1 per cent (wt/vol) agar solution
- 70 per cent (vol/vol) ethanol
- Sodium hypochlorite solution containing 1 per cent available chlorine and 0.02 per cent (vol/vol) Tween 20
- Distilled water
- *MS medium:* 1× MS plant salt mixture (Wako Pure Chemical Industries, Osaka, Japan), 1× Gamborg's vitamin solution (Sigma-Aldrich, St. Louis, MO, USA), 1 per cent (wt/vol) sucrose, 0.05 per cent (wt/vol) MES, and pH 5.7 adjusted with 1 N KOH
- Bacto agar (Becton, Dickinson and Company, Franklin Lakes, NJ, USA) (recommended)
- LB agar plate
- Liquid LB

- Glycerol
- *Transformation buffer:* 1/2× MS plant salt mixture, 1× Gamborg's vitamin solution, 5 per cent (wt/vol) sucrose, and pH 5.7, adjusted with 1 N KOH
- 5 per cent (wt/vol) sucrose solution
- Silwet L-77
- 6-Benzylaminopurine (BAP) (final concentration 0.01 μg ml^{-1})
- Claforan
- Kanamycin (final concentration 30 μg ml^{-1})
- Hygromycin (final concentration 20 μg ml^{-1})
- Bialaphos (final concentration 7.5 μg ml^{-1})
- Peat moss (Soil Mix, Sakata Seed Corp., Yokohama, Japan)
- Expanded vermiculite granules

Equipments

- Growth chamber
- Plant pot (3-inch)
- Conical tube (15 ml)
- Eppendorf tubes (2 ml)

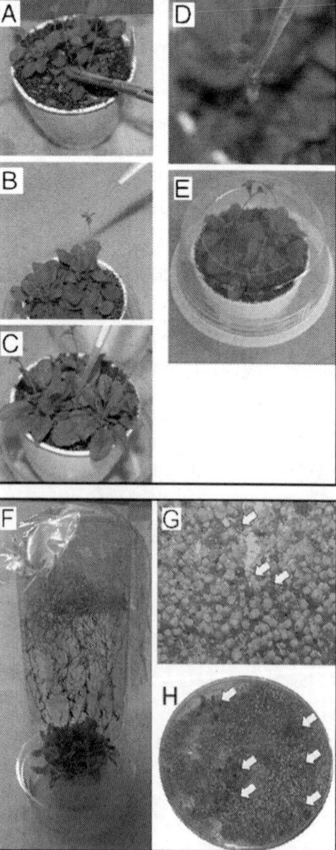

Fig. Floral Inoculating Transformation of *Arabidopsisthaliana*.

(A) Clipping Primary Bolts. (B, C, and D) Using a Micropipette, Inoculate Flower Buds with 5 μl of *Agrobacterium* when Plants have Just Started to Flower after Clipping Primary Bolts. (E) Place Inoculated Plants Under a Dome or Cover for 16–24 hrs to Maintain High Humidity. (F) Remove the Cover and Grow the Plants in a Greenhouse or Growth Chamber Until Maturity. (G, H) Screening of Putative Transformed *Arabidopsis* Plants. G: 10 days, H: 21 days. Arrows Indicate Putative Transformed *Arabidopsis* Plants.

- Grow A. *thaliana* plants. (Note: Plant health is an important factor. Healthy A. *thaliana* plants should be grown until they are flowering.) There are two different procedures: standard (A) and quick (B) (Zhang *et al.*, 2006). We generally use the quick procedure, which is useful for rare seeds and seeds with low germination frequency. It is also used to retransform a transgenic line with a second construct.
  - *1.a.Standard procedure (A):* Suspend seeds in 0.1 per cent (wt/vol) agar solution and keep in darkness for 2–4 days at 4°C to break dormancy. Spread seeds on wet soil (a mixture of peat moss and expanded vermiculite granules at a 1:2 ratio) in a 3-inch pot and grow under long-day conditions (16-hr light/8-hr dark) at 22°C. Thin to three seedlings per pot. Do not cover with a bridal veil, window screen, or cheesecloth.
  - *1.b.Quick procedure (B):* Sterilise seeds by treatment with 70 per cent (vol/vol) ethanol for 1 min then immerse in sodium hypochlorite solution containing 1 per cent available chlorine and 0.02 per cent (vol/vol) Tween 20 for 7 min. Wash seeds five times with sterile distilled water. Place seeds on MS medium containing 0.8 per cent (wt/vol) Bacto agar. Keep seeds in darkness for 2–4 days at 4 ºC to break dormancy. Grow under long-day conditions (16-hr light/ 8-hr dark) for 3 weeks at 22°C. Transfer to pots per Step 1a. Do not cover with a bridal veil, window screen, or cheesecloth.
- Clip primary bolts to encourage proliferation of secondary bolts (Figure A). Plants will be ready approximately 4–6 days after clipping.
- Prepare the *Agrobacteriums* train carrying the gene of interest. Spread a single *Agrobacterium* colony on an LB agar plate with suitable antibiotics. Incubate the culture at 28°C for two days.
- Use feeder culture to inoculate a 2-ml liquid culture in LB with suitable antibiotics to select for the binary plasmid in a 15-ml Conical tube at 28°C for 16–24 hrs. Mid-log cells or a freshly saturated culture (Clough and Bent 1998) can be used. (Optional: If needed, keep 500 μl of *Agrobacterium* culture in a 25 per cent (vol/vol) glycerol stock at –80°C.)
- Spin down 1.5 ml of the *Agrobacterium* cell suspension in 2-ml Eppendorf tubes and resuspend in 1 ml transformation buffer. OD_{600} value adjustment is not required. Each small pot containing three

plants requires approximately 150 μl of culture. (Optional: 5 per cent (wt/vol) sucrose solution may be used instead of transformation buffer.) Just before inoculation, add Silwet L-77 to a concentration of 0.02 per cent (vol/vol) and immediately mix well. (Optional: If using transformation buffer, add 0.01 μg ml^{-1} BAP just before transformation.)

- Apply 5 μl of *Agrobacterium* inoculum to the flower buds (Figures B, C, and D), inoculating each plant with a total of 30–50 μl of inoculum.
- Place inoculated plants under a dome or cover for 16–24 hrs to maintain high humidity (Figure E). Avoid excessive exposure to light. (Optional: For higher rates of transformation, inoculate newly forming flower buds with *Agrobacterium* 2–3 times at 7-day intervals.)
- Water and grow plants normally, tying up loose bolts with wax paper, tape, stakes, twist-ties, or other means. Stop watering as seeds become mature (Figure F).
- Harvest dry seeds. Though transformants are usually independent, independence can be guaranteed if seeds come from separate plants.
- Surface-sterilise seeds by immersion in 70 per cent (vol/vol) ethanol for 1 min, followed by immersion in sodium hypochlorite solution containing 1 per cent available chlorine and 0.02 per cent (vol/vol) Tween 20 for 10 min. Then, wash seeds five times with sterile distilled water.

  To select for transformed plants, resuspend liquid-sterilised seeds in approximately 8 ml of 0.1 per cent (wt/vol) agar solution containing 2 mg ml^{-1}Claforan. Sow seeds per Step 1b in MS medium containing 0.8 per cent Bacto agar and appropriate antibiotics or herbicide selective markers at the following concentrations: kanamycin (final concentration 30 μg ml^{-1}), hygromycin (20 μg ml^{-1}), and bialaphos (7.5 μg ml^{-1}). Claforan is necessary for *Agrobacterium* decontamination (Figures G and H).
- Transplant putative transformants to soil per Step 1a. Grow, test, and use.

Showing of Transgenic Plants by Pcr

Transgenes can be detected by plant genome DNA analysis with PCR. Although transgenes can be distinguished from their surrounding host plant genome, their presence should be determined by DNA sequencing. PCR-based transgene detection is a simple and highly sensitive process. Subsequent PCR tests are assessed by agarose gel electrophoresis, and results are visualised by the presence or absence of the appropriately sized DNA fragment. If PCR shows a positive result, the transgene may be present. Transgene presence is confirmed by incorporating it into the genome by DNA sequencing. In contrast, a negative PCR result implies that the transgene is not present.

GENETIC TRANSFORMATION

Tissue culture techniques have become an important part in helping the success of plant genetic engineering (gene transfer). For example, bacterial gene transfer (such as cry genes from Bacillus thuringiensis) into the plant cells will be expressed after transgeniknya achieved plant regeneration.

THE PRODUCTION OF SECONDARY METABOLITES, COMPOUNDS

Plant cell culture can also be used to produce biochemical compounds (secondary metabolites) such as alkaloids, terpenoids, phenyl etc. propanoid. This technology is now available in industrial scale. For example, the commercial production of compounds "shikonin" from Lithospermum erythrorhizon cell culture. Plant cells can be grown in isolation from intact plants in tissue culture systems. The cells have the characteristics of callus cells, rather than other plant cell types. These are the cells that appear on cut surfaces when a plant is wounded and which gradually cover and seal the damaged area.

Pieces of plant tissue will slowly divide and grow into a colourless mass of cells if they are kept in special conditions.

These are:

- Initiated from the most appropriate plant tissue for the particular plant variety
- Presence of a high concentration of auxin and cytokinin growth regulators in the growth media
- A growth medium containing organic and inorganic compounds to sustain the cells
- Aseptic conditions during culture to exclude competition from microorganisms

The plant cells can grow on a solid surface as friable, pale-brown lumps (called callus), or as individual or small clusters of cells in a liquid medium called a suspension culture. These cells can be maintained indefinitely provided they are sub-cultured regularly into fresh growth medium.

Tissue culture cells generally lack the distinctive features of most plant cells. They have a small vacuole, lack chloroplasts and photosynthetic pathways and the structural or chemical features that distinguish so many cell types within the intact plant are absent. They are most similar to the undifferentiated cells found in meristematic regions which become fated to develop into each cell type as the plant grows. Tissue cultured cells can also be induced to re-differentiate into whole plants by alterations to the growth media.

Plant tissue cultures can be initiated from almost any part of a plant. The physiological state of the plant does have an influence on its response to attempts to initiate tissue culture. The parent plant must be healthy and free from obvious signs of disease or decay. The source, termed explant, may be dictated by the reason for carrying out the tissue culture. Younger tissue contains a higher proportion of actively dividing cells and is more responsive to a callus initiation programme.

The plants themselves must be actively growing, and not about to enter a period of dormancy. The exact conditions required to initiate and sustain plant cells in culture, or to regenerate intact plants from cultured cells, are different for each plant species. Each variety of a species will often have a particular set of cultural requirements. Despite all the knowledge that has been obtained about plant tissue culture during the twentieth century, these conditions have to be identified for each variety through experimentation.

Table. Secondary Metabolites Produced in High Levels by Plant Cell Cultures

Compound	Plant Species	Yields (% dry wt)	Culture	
		Culture	Plant	Type*
Shikonin	Lithospermum erythrorhizon	20	1.5	s
Ginsenoside	Panax ginseng	27	4.5	c
Anthraquinones	Morinda citrifolia	18	0.3	s
Ajmalicine	Catharanthus roseus	1.0	0.3	s
Rosmarinic acid	Coleus blumeii	15	3	s
Ubiquinone-10	Nicotiana tabacum	0.036	0.003	s
Diosgenin	Dioscorea deltoides	2	2	s
Benzylisoquinoline	Coptis japonica	11	5 - 10	s
Alkaloids				
Berberine	Thalictrum minor	10	0.01	s
Berberine	Coptis japonica	10	2 - 4	s
Anthraquinones	Galium verum	5.4	1.2	s
Anthraquinones	Galium aparine	3.8	0.2	s
Nicotine	Nicotiana tabacum	3.4	2.0	c
Bisoclaurine	Stephania cepharantha	2.3	0.8	s
Tripdiolide	Tripteryqium wilfordii	0.05	0.001	s

* s = suspension; c = callus

As seen in many reviews, studies on the production of plant metabolites by callus and cell suspension cultures have been carried out on an increasing scale since the end of the 1950's. The large scale cultivation of tobacco and a variety of vegetable cells was examined from the late 1950's to early 1960's by Tulecke and Nickell at Pfizer Inc., Mandels et at. at the Natick Laboratories in the U.S. Army, Street *et al.* at the University of Leicester and Martin *et al.* at the National Research Council of Canada. Their results stimulated more recent studies on the industrial application of plant cell culture in many countries.

Since Japan has a highly developed fermentation technology, many industrial companies, in collaboration with some university groups, have tried to apply this technology for the commercial production of useful compounds. The Japan Tobacco Inc.'s interest involved around mass-production of tobacco cells as raw materials of cigarettes; the company established 20 kL fermentors which were the largest for plant cells in 1970's. Meiji Seika in Japan also elucidated the fundamentals of production of Panax ginseng cells in large volumes.

Researchers reported that cultured ginseng cells stimulated physiological activities in animals in a similar fashion as elicited by native ginseng roots. The work was followed by Nitto Denko Co. which has been manufacturing cell mass of ginseng commercially. The cells are used as health foods in Japan. Researchers of Kyowa Hakko conducted extensive pharmacological screening of numerous cell cultures and found various novel products of great interest, including plasmin inhibitory proteins in Scopolia japonica cells and plant virus inhibitors in cultured cells of Phytolacca americana and other species.

The virus inhibitors of P. americana are now being studied by many research groups in the world because of their activity against AIDS and other animal viruses. Other firms such as Ajinomoto and Nippon Shin-yaku also made efforts to increase the level of accumulation of alkaloids, steroids and other secondary products in cultured cells.

Groups in Germany outlined very interesting approaches to industrial application in the meeting held in 1976 at Munich, and their excellent results encouraged researchers in other countries. For example, Zenk and his colleagues presented an impressive paper in which they successfully selected high-alkaloid producing cell lines of Catharanthus roseus using a method similar to the microbial mono-colony isolation technique. A number of laboratories in industries and universities followed Zenk's approach, and in fact, some researchers could increase significantly the level of secondary products such as ubiquinone-10, biotin, and various plant pigments produced by cell cultures.

In 1982, the 5th International Congress of Plant Tissue and Cell Cultures was held in Japan and about papers presented there related to production of secondary metabolites in cultured cells and several papers seemed to be commercially promising such as production of shikonin by Fujita *et al.*) of Mitsui Petrochemical and that of several antitumor compounds by Misawa *et al.*

At subsequent International Congress of Plant Tissue and Cell Cultures in Minneapolis in 1986 and that in Amsterdam in 1990 as well as other meetings such as the Meeting of Primary and Secondary Metabolism of Plant Cell Cultures held in Giessen, Germany and the 4th and International Congress on Phytotherapy held in Munich, Germany in September, 1992, many compounds were shown to be accumulated by plant cell cultures and many different strategies were presented to increase their productivity. Means for production of those compounds include not only de novo synthesis but also biotransformation processes. A biotransformation process to produce β-methyl digoxin using Digitalis lanata cells studied by Reinhard and Alfermann in Germany was evaluated by Boehringer Mannheim Co. using 4 kL bioreactors although it has not yet been commercialized. A combination of a plant cell culture process and a simple chemical coupling reaction was invented by a Canadian company, Allelix, to manufacture vinblastine as a commercially feasible process. The technology is now being studied by Mitsui Petrochemical in Japan for comm-ercialization.

In spite of remarkable advances in plant cell culture technology, the production cost of metabolites is still high, as estimated by Zenk and by Goldstein. The former calculated that the cost would be U.S. $500 per kg of isolated product when it was accumulated in 1 g/L within a period of 15 days in 100 m batch culture. It is true that the cost has decreased since his estimation in 1974, and the producing ability of 1 g/L for 15 days is achievable in the case of some compounds such as rosmarinic acid, but it is still too expensive to produce food stuffs, food additives or pharmaceuticals which can be more easily produced by alternative ways such as chemical synthesis, extraction from plants or by microbial fermentation.

Zenk stated that industrial plant cell culture techniques would be introduced only if the plant product under consideration is produced at a price equal to or preferably lower than the field-produced product. The only factor determining the industrial realization of plant cell culture is the price by which a given product can be produced. A very detailed assessment of the costs entailed in theoretical processes has been made by Goldstein *et al.*

Four cases were considered:

A (Base state of technology)
Final biomass concentration = 200 g fresh wt./L
One day doubling time, biomass = 0.693/day
One day doubling time, product = 0.693/day
Product concentration = 0.05 per cent fresh weight
Recycled biomass = 50 per cent

B *As for A, but*:
One day doubling time, product = 6.93/day (10 times increase)
Recycled biomass = 90 per cent

C *As for B, but:*
Final biomass concentration = 500 g fresh wt./L (2.5 times increase)

D *As for C, but*:
Product concentration = 0.5 per cent fresh weight (10 times increase).

The final case (D) would translate into a production of 17.5 g product/L/day, which is well in excess of levels achieved with current technology. The rate of product formation is the major obstacle, but this could be compensated for by increased concentration of product. The calculated manufacturing costs and profit margin for each case assuming production of 104, 105 or 106 kg/yr are shown in Tables. Note that these figures reflect 1980 prices and will have increased since then.

Table. Manufacturing Costs for Cases A - D

Production Level (kg/yr)	10^4	10^5	10^6
	Cost ($/kg)		
Case:			
A.	551	241	106
B.	323	142	61
C.	165	72	32
D.	92	40	18

Table. Profitability of Cases A - D The Selling Price ($) and Profit (%) are Given for each of the Three Annual Production Levels.

Production Level (kg/yr)	10^4		10^5		10^6	
	$	%	$	%	$	%
Case:						
A.	1,608	5.7	683	14.7	290	13.8
B.	992	17.5	421	16.3	177	15.3
C.	528	18.7	223	17.6	94	16.6
D.	233	10.6	100	9.8	43	9.1

The selling price was calculated to take into account a 5-year payment on capital and a tax rate of 50 per cent. The formula used was: Selling price/kg = 2 x manufacturing cost/kg + 0.4 (total invested capital/kg) = depreciation/kg.

In the case of scenario (D) with production of 10 kg/yr the selling price could be as low as $43/kg and the manufacturing cost would only be $18/kg. This assessment makes the future of commercial plant cell culture fermentation appear quite viable, but the crucial problems to be overcome are enhancing product levels, recycling the biomass (*e.g.* through immobilization) and raising the biomass concentration. Some encouraging advances in recent years suggest that these problems may be solved eventually, *e.g.* shikonin can be produced at a level of 23 per cent of the dry weight (roughly 2.3 per cent or more of fresh weight) and 4 g/L; rosmarinic acid can be produced at a rate of 1 g/L/ day in Coleus blumei with a final concentration of 27 per cent of the dry weight. A biomass concentration of 40 g dry wt./L was achieved.

Batch fermentation processes are capital intensive, requiring a high initial investment and sophisticated technology. In comparison with microbial cultures, doubling times are long and product yields are low. Furthermore, since production is often not growth-linked, there are many cases where two stage production is necessary, thus increasing costs markedly. Table shows estimated costs for a hypothetical phytochemical produced by plant cell culture fermentation.

Table. Batch Fermentation Costs for a Hypothetical Plant Product as a Function of Product Concentration

Product Concentration (%)	0.1	1.0	10.0
Fixed Capital Investment ($M)	340	84	21
Total Production Cost (%/kg)	5,900	1,045	228
Assumptions:	Annual production = 20,000 kg Cell doubling time = 60 hr Batch cycle time = 15 d Cell intensity = 20 g/L		

The total production cost is the sum of direct and indirect costs, including raw material, labour, utilities, capital eq;uipment depreciation over a 10 year period, maintenance, tax and insurance. Indirect costs were taken as 15 per cent of capital investment

It has been calculated that if rose oil were produced in cell cultures, a fermentation process with a 20 per cent share of the market would yield a net profit of $0.5 M/yr. This assumes a rose oil concentration of 10 per cent dry weight and a cell density of 20 g dry weight/L. Fowler suggests that plant products valued at $250-1,500/kg are acceptable production targets for an economically viable process. An initial market penetration of the world or U.S. market has been recommended at between 10 and 20 per cent. This will depend on the size of the market and whether a plant cell fermentation process opens up new market.

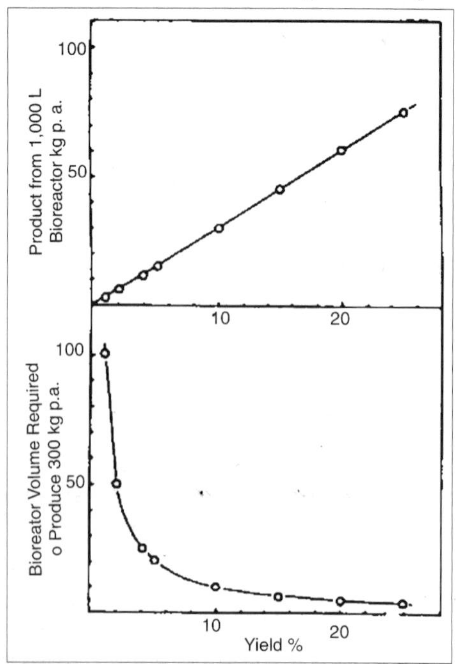

Fig. The Effect of Run Time on Bioreactor Volume and the Production from a 1,000 L Bioreactor

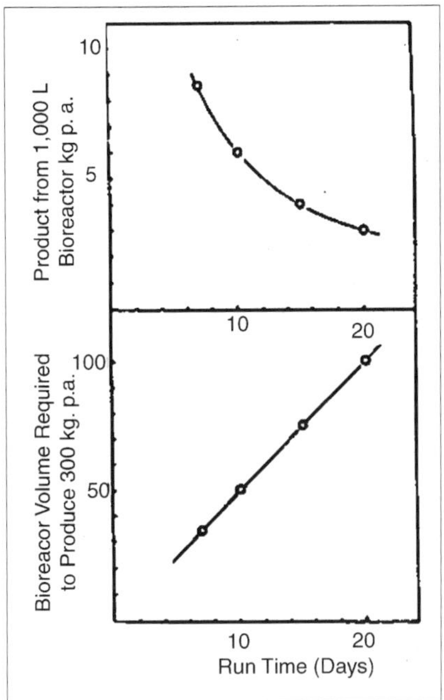

Fig. The Effect of Yield on Bioreactor Volume and the Production from a 1,000 L. Bioreactor

Scragg has calculated the effect of run time on productivity. Assuming a product yield of 1 per cent dry weight, biomass yield of 20 g dry weight/L and a bioreactor volume of 1,000 L. A 300 day year was estimated, since time would be required for maintenance, contaminated runs, etc. Fig. shows that as productivity decreases the requisite run time increases. Similarly, as the yield of product increases the reactor volume required decreases expotentially. Under the conditions described above, a 100,000 L fermentor would be necessary to produce 300 kg/yr.

The most detailed assessment available in the literature for the production of a specific product in plant cell cultures is concerned with the production of ajmalicine from Catharanthus roseus cell cultures. In this case, a 20 per cent market penetration was assumed, ie. 800 kg/yr and production was based on the use of a two-stage batch culture process. Stage two parameters include specific productivity of 0.26 mg/g/day (based on final dry weight), final product concentration of 0.06 per cent dry weight and a maximum fresh weight concentration of 160 g/L.

It was estimated that the cost of production would be $3,215/kg ajmalicine and this compares with $619/kg ($0.70/lb dry biomass) from the intact plant. Clearly, this 5-fold increase in costs is unacceptable from a commercial point of view. This example serves to illustrate the need to chose products that are particularly expensive to produce in the field. Shikonin is a good example; in that the long growing period and strict climatic requirements mean that the cost of the plant raw material is high $6.80/lb. It has been suggested that a quick way to assess the attractiveness of a cell culture method vs. conventional agriculture, is to calculate a specific biosynthesis rate "based on the final dry weight and the total time of fermentation or land occupation". With shikonin there is a 830-fold increase in plant cell culture, whilst with ajmalicine from C. roseus there was only a 24-fold improvement.

HISTORY

The use of tissue culture techniques initiated by Gottlieb Haberlandt in 1902 in an attempt mengkulturkan hair cells of the leaf mesophyll tissue monocot plant. But the effort failed because the cells do not have cleavage, it was alleged failures because they do not use growth regulator substances needed for cell division, proliferation and induction of the embryo. In the year 1904, Hannig planting embryos isolated from several plant crucifers. In 1922, Knudson and separately each Robbin conduct investment business and culture of orchid seedlings root tip. After the 1920s, the discovery and development of tissue culture techniques continues.

DEVELOPMENT OF GENETIC AND METABOLIC

The vitro techniques were developed initially to demonstrate the totipotency of plant cells predicted by Haberlandt in 1902. Totipotency is the ability of a plant cell to perform all the functions of development, which are characteristic of zygote, *i.e.*, ability to develop into a complete plant. In 1902, Haberlandt reported culture of isolated single palisade cells from leaves in Knop's salt solution enriched with sucrose. The cells remained alive for up to 1 month, increased in size, accumulated starch but failed to divide. Efforts to demonstrate totipotency led to the development of techniques for cultivation of plant cells under defined conditions. Haploid plants from pollen grains were first produced by Maheshwari and Guha in 1964 by culturing anthers of Datura. This marked the beginning of anther culture or pollen culture for the production of haploid plants.

The technique was further developed by many workers, more notably by JP. Nitch, C. Nitch and coworkers. These workers showed that isolated microspores of tobacco produce complete plants. Plant protoplasts are naked cells from which cell wall has been removed. In 1960, Cocking produced large quantities of protoplasts by using cell wall degrading enzymes.

The techniques of protoplast production have now been considerably refined. It is now possible to regenerate whole plants from protoplasts and also to fuse protoplasts of different plant species. In 1972, Carlson and coworkers produced the first somatic hybrid plant by fusing the protoplasts of Nicotiana glauca and N. langsdorfii. Since then many divergent somatic hybrids have been produced. A successful establishment of callus cultures depended on the discovery during mid-thirties of IAA (idole-3-acetic acid), the endogenous auxin, and of the role of B vitamins in plant growth and in root cultures. The first continuously growing callus cultures were established from cambium tissue in 1939 independently by Gautheret, White and Nobecourt.

The subsequent discovery of kinetin by Miller and coworkers in 1955 enabled the initiation of callus cultures from differentiated tissues. Shoot bud differentiation from tobacco pith tissues cultured in vitro was reported by Skoog in 1944, and in 1957 Skoog and Miller proposed that root-shoot differentiation in this system was regulated by auxin-cytokinin ratio.

The first plant from a mature plant cell was regenerated by Braun in 1959. Development of somatic embryos was first reported in 1958- 1959 from carrot tissues independently by Reinert and Steward. Thus within a brief period, the tissue culture techniques have made a great progress. From the sole objective of demonstrating the totipotency of differentiated plant -cells, the technique now finds application in both basic and applied researches in a number of-fields of enquiry.

TYPES

Tissue culture consists of growing plants cells as relatively on organized masses of cells on an agar medium (callus culture) or as a suspension of free cells and small cell masses in a liquid medium (suspension culture). Tissue culture is used for vegetative multiplication of many species and in some cases for recovery of virus-free plants. It has potential application in production of somatic hybrids, organelle and cytoplasm transfer, genetic transformation and germplasm storage through freeze-preservation.

CLONAL PROPAGATION

Tissue culture is well suited for quick vegetative propagation of plant species. It is used for asexual propagation in many species of fruit and timber trees and also used for obtaining disease free and virus-free plants. The major difficulty in the use of this technique in clonal multiplication is the occurrence of genetic variation among the regenerated plants. This problem can be reduced to a large extent by using young tissue cultures, preferably during the first few subcultures.

MUTANT ISOLATION

Biochemical mutants are far more easily isolated from cell cultures than from whole plant populations. This is because a large number of cells, 10^6-10^9, can be easily and effectively screened for biochemical mutant cells. Biochemical mutants could be selected for disease resistance, improvement of nutritional quality, adaptation of plants to stress conditions, *e.g.* saline soils, and to increase the biosynthesis of plant products used for medicinal or industrial purposes.

SOMACLONAL VARIATION

Plants regenerated from tissue and cell cultures show heritable variation for both qualitative and quantitative traits; such a variation is known as somaclonal variation. Somalconal variation has been described in sugarcane, potato, tomato etc. Some variants are obtained in homozygous condition in the plants regenerated from the cells in vitro (R_0 generation), but most variants are recovered in the selfed progeny of the tissue culture-regenerated plants (R_1 generation).

Somaclonal variation most likely arises as a result of chromosome structural changes, *e.g.*, small deletions and duplications, gene mutations, plasma gene mutations, mitotic crossing over and possibly, transposons. Somaclonal variation may be profitably utilized in crop improvement since it reduces the time required for releasing the new variety by at least two years as compared to mutation breeding and by three years in comparison to back cross method of gene transfer.

AMINO ACID RESISTANT MUTANTS

Cereal grains are deficient in lysine; maize (Zea maize) is also deficient in trytophan, while wheat (T.aestivum) and rice (O.sativa) are deficient in threonine. Pulses are deficient in methionine and trytophan. Amino acid analogue-resistant cells may be expected to show a relatively higher concentration of that particular amino acid. For *e.g.*, carrot (D.carota) and tobacco (N.tabacum) cell lines resistant to trytophan analogue 5-methyl trytophan show a 10-27-fold increase in the level of trytophan. Similarly, rice cells resistant to lysine analogue 5-(B-aminoethyl)-cysteine, show much higher levels of lysine. This technique may prove useful in the development of crop varieties with a better-balanced amino acid content.

DISEASE RESISTANT MUTANTS

Many pathogenic bacteria produce toxins that ae toxic to plant cells. Plant cell cultures may be exposed to lethal concentrations of these toxins and resistant clones isolated. Plants regenerated from these resistant clones would be resistant to the disease producing pathogen. This technique should be applicable to all the pathogens, which produce the disease through the action of toxin. An *e.g.*, an application is in the case of wildfire disease of tobacco (N.tabacum) produced by Pseudomonos tabaci.

Tobacco cells resistant to methionine sulfoximine, which is similar to the toxin produced by the pathogen, were isolated. Plants regenerated from these clones were resistant to wildfire disease, although to a somewhat lesser degree. The technique can be applied to those cases only where the disease is the result of a toxin produced by the pathogen. But many of the pathogens do not seem to produce a toxin, or the toxin does not appear to be the primary cause of the disease.

Stress Resistant and Other Mutants

Plant cells resistant to 4-5 times the normally toxic salt (NaCl) concentration have been isolated. Attempts to insolate such cells are being made. Similarly, attempts are being made to isolate clones that would produce more substances of medicinal or industrial value.

SOMATIC HYBRIDIZATION

Protoplasts can be isolated from almost every plant species and cultured to produce callus. Protoplasts of two different species may be fused with the help of polyethylene glycol.

GENETIC TRANSFORMATION

There is some evidence that gene transfer may be achieved by feeding cells with DNA in case of eukaryotes, such as, Drosophila, Neurospora, cultured mammalian cells and in some plants. Genetic changes may be brought about by DNA or by radiation-killed pollen grains. This raises the possibility of genetic modification of plant cells with the help of both homologous (from the same species) and heterologous (from a different species) DNA. It is also proposed that DNA plant viruses, such as cauliflower (B.oleracea) mosic virus and potato leaf roll virus, plasmids (*e.g.*, Ti plasmid of Agrobacterium) and transposons, may be used as the carriers of genes for genetic modification of plant cells.

Organelle Transfer

In some cases, it may be desirable to transfer only organelles or the cytoplasm into a new genetic background. This may be achieved through the use of plant protoplasts. Chloroplasts have been transferred, and other organelles including nucleus may be transferred.

Germplasm Conservation

Tissue cultures may be frozen and stored in liquid nitrogen at –1960C for long-term storage of germplasm. This would be of great value in the conservation of germplasm of those crops which normally do not produce seeds, *e.g.* root and tuber crops, or where it may not be desirable to store seeds. For freeze-preservation, the cells are cooled at a slow rate and are then transferred to liquid nitrogen for storage. Thawing of the cells must be very rapid for increased survival. A cryoprotectant, such as dimethylsufoxide (DMSO), is used to protect the cells from injury due to freezing and thawing. The technique of freeze-preservation *i.e.*, crybiology, of plant cells is still in the developing stages.

ACHIEVEMENTS AND FUTURE PROSPECTS

Tissue culture techniques are being exploited to enhance crop production and to aid crop improvement efforts. Faster clonal multiplication is being exploited on commercial scale for many horticultural species *e.g.* oil palm, mentha, roses, carnation etc. Tissue cultured somatic tissues are now routinely being used for conservation of those species whose seeds are recalcitrant or ones which do not produce seed at all. Embryo culture has helped in rescuing hybrid embryos enabling the recovery of many interspecific hybrids and haploid plants. Shoot tip (meristem) culture plays a vital which is of great importance in germplasm exchange, and the development of serological techniques for the detection of viruses in plant materials is a great help to the efforts in this direction.

COMMERCIALLY GROWN OF TRANSGENIC PLANTS

Some of the commercially grown transgenic plants in developed countries are: "Roundup Ready" soybean, 'Freedom II squash', 'High- lauric' rapeseed (canola), 'Flavr Savr' and 'Endless Summer' tomatoes. During 1995, full registration was granted to genetically engineered Bt gene containing insect resistant 'New Leaf' (potato), 'Maximizer' (corn), 'BollGard' (cotton) in USA. Some of the traits introduced in these transgenic plants are as follows:

STRESS TOLERANCE

Biotechnology strategies are being developed to overcome problems caused due to biotic stresses (viral, bacterial infections, pests and weeds) and abiotic stresses (physical actors such as temperature, humidity, salinity, etc.).

Abiotic Stress Tolerance

The plants show their abiotic stress response reactions by the production of stress related osmolytes like sugars (*e.g.*, trehalose and fructans), sugar alcohols (*e.g.*, mannitol), amino acids (*e.g.*, proline, glycine, betaine) and certain proteins (*e.g.*, antifreeze proteins). Transgenic plants have been produced which over express the genes for one or more of the above mentioned compounds. Such plants show increased tolerance to environmental stresses. Resistance to abiotic stresses includes stress induced by herbicides, temperature (heat, chilling, freezing), drought, salinity, ozone and intense light. These environmental stresses result in the destruction, deterioration of crop plants which leads to low crop productivity. Several strategies have been used and developed to build resistance in the plants against these stresses.

Herbicide Tolerance

Weeds are unwanted plants which decrease the crop yields and by competing with crop plants for light, water and nutrients. Several biotechnological strategies for weed control are being used, *e.g.*, the over-production of herbicide target enzyme (usually in the chloroplast) in the plant which makes the plant insensitive to the herbicide. This is done by the introduction of a modified gene that encodes for a resistant form of the enzyme targeted by the herbicide in weeds and crop plants. Roundup Ready crop plants tolerant to herbicide-Roundup, is already being used commercially. The biological manipulations using genetic engineering to develop herbicide resistant plants are:

- Over-expression of the target protein by integrating multiple copies of the gene or by using a strong promoter.,
- Enhancing the plant detoxification system which helps in reducing the effect of herbicide.,
- Detoxifying the herbicide by using a foreign gene., and
- Modification of the target protein by mutation.

Some of the examples are: Glyphosate resistance - Glyphosate is a glycine derivative and is a herbicide which is found to be effective against the 76 of the world's worst 78 weeds. It kills the plant by being the competitive inhibitor of the enzyme 5-enoyl-pyruvylshikimate 3- phosphate synthase (EPSPS) in the shikimic acid pathway. Due to it's structural similarity with the substrate phosphoenol pyruvate, glyphosate binds more tightly with EPSPS and thus blocks the shikimic acid pathway.

Certain strategies were used to provide glyphosate resistance to plants:

- It was found that EPSPS gene was overexpressed in Petunia due to gene amplification. EPSPS gene was isolated from Petuniaand introduced in to the other plants. These plants could tolerate glyphosate at a dose of 2- 4 times higher than that required to kill wild type plants.
- By using mutant EPSPS genes- A single base substitution from C to T resulted in the change of an amino acid from proline to serine in EPSPS. The modified enzyme cannot bind to glyphosate and thus provides resistance.
- The detoxification of glyphosate by introducing the gene (isolated from soil organism- Ochrobactrum anthropi) encoding for glyphosate oxidase into crop plants. The enzyme glyphosate oxidase converts glyphosate to glyoxylate and aminomethylphosponic acid. The transgenic plants exhibited very good glyphosate resistance in the field.

Phosphinothricin is a broad spectrum herbicide and is effective against broad-leafed weeds. It acts as a competitive inhibitor of the enzyme glutamine synthase which results in the inhibition of the enzyme glutamine synthase and accumulation of ammonia and finally the death of the plant. The disturbance in the glutamine synthesis also inhibits the photosynthetic activity. The enzyme phosphinothricin acetyl transferase (which was first observed in Streptomyces sp in natural detoxifying mechanism against phosphinothricin) acetylates phosphinothricin, and thus inactivates the herbicide. The gene encoding for phosphinothricin acetyl transferase (bar gene) was introduced in transgenic maize and oil seed rape to provide resistance against phosphinothricin.

Other Abiotic Stresses

The abiotic stresses due to temperature, drought, and salinity are collectively also known as water deficit stresses. The plants produce osmolytes or osmoprotectants to overcome the osmotic stress. The attempts are on to use genetic engineering strategies to increase the production of osmoprotectants in the plants. The biosynthetic pathways for the production of many osmoprotectants have been established and genes coding the key enzymes have been isolated. *E.g.* Glycine betaine is a cellular osmolyte which is produced by the participation of a number of key enzymes like choline dehydrogenase, choline monooxygenase, etc.

The choline oxidase gene from Arthrobacter sp. was used to produce transgenic rice with high levels of glycine betaine giving tolerance against water deficit stress. Scientists also developed cold-tolerant genes (around 20) in Arabidopsis when this plant was gradually exposed to slowly declining temperature. By introducing the coordinating gene (it encodes a protein which acts as transcription factor for regulating the expression of cold tolerant genes), expression of cold tolerant genes was triggered giving protection to the plants against the cold temperatures.

Insect Resistance

A variety of insects, mites and nematodes significantly reduce the yield and quality of the crop plants. The conventional method is to use synthetic pesticides, which also have severe effects on human health and environment. The transgenic technology uses an innovative and eco-friendly method to improve pest control management. About 40 genes obtained from microorganisms of higher plants and animals have been used to provide insect resistance in crop plants.

The first genes available for genetic engineering of crop plants for pest resistance were Cry genes (popularly known as Bt genes) from a *bacterium Bacillus thuringiensis*. These are specific to particular group of insect pests, and are not harmful to other useful insects like butter flies and silk worms. Transgenic crops with Bt genes (*e.g.*, cotton, rice, maize, potato, tomato, brinjal, cauliflower, cabbage, etc.) have been developed. This has proved to be an effective way of controlling the insect pests and has reduced the pesticide use.

The most notable example is Bt cotton (which contains CrylAc gene) that is resistant to a notorious insect pest Bollworm *(Helicoperpa armigera)*.. There are certain other insect resistant genes from other microorganisms which have been used for this purpose. Isopentenyl transferase gene from Agrobacterium tumefaciens has been introduced into tobacco and tomato. The transenic plants with this transgene were found to reduce the leaf consumption by tobacco hornworm and decrease the survival of peach potato aphid.

Certain genes from higher plants were also found to result in the synthesis of products possessing insecticidal activity. One of the examples is the Cowpea trypsin inhibitor gene (CpTi) which was introduced into tobacco, potato, and oilseed rape for developing transgenic plants. Earlier it was observed that the wild species of cowpea plants growing in Africa were resistant to attack by a wide range of insects. It was observed that the insecticidal protein was a trypsin inhibitor that was capable of destroying insects belonging to the orders Lepidoptera, Orthaptera, etc. Cowpea trypsin inhibitor (CpTi) has no effect on mammalian trypsin, hence it is non-toxic to mammals.

Virus Resistance

There are several strategies for engineering plants for viral resistance, and these utilises the genes from virus itself (*e.g.*, the viral coat protein gene). The virus-derived resistance has given promising results in a number of crop plants such as tobacco, tomato, potato, alfalfa, and papaya. The induction of virus resistance is done by employing virus-encoded genes-virus coat proteins, movement proteins, transmission proteins, satellite RNa, antisense RNAs, and ribozymes. The virus coat protein-mediated approach is the most successful one to provide virus resistance to plants. It was in 1986, transgenic tobacco plants expressing tobacco mosaic virus (TMV) coat protein gene were first developed. These plants exhibited high levels of resistance to TMV. The transgenic plant providing coat protein-mediated resistance to virus are rice, potato, peanut, sugar beet, alfalfa, etc. The viruses that have been used include alfalfa mosaic virus (AIMV), cucumber mosaic virus (CMV), potato virus X (PVX), potato virus Y (PVY) etc.

RESISTANCE AGAINST FUNGAL AND BACTERIAL INFECTIONS

As a defence strategy against the invading pathogens (fungi and bacteria) the plants accumulate low molecular weight proteins which are collectively known as pathogenesis-related (PR) proteins. Several transgenic crop plants with increased resistance to fungal pathogens are being raised with genes coding for the different compounds. One of the examples is the Glucanase enzyme that degrades the cell wall of many fungi. The most widely used glucanase is beta-1,4-glucanase. The gene encoding for beta-1,4 glucanase has been isolated from barley, introduced, and expressed in transgenic tobacco plants. This gene provided good protection against soil-borne fungal pathogen Rhizoctonia solani. Lysozyme degrades chitin and peptidoglycan of cell wall, and in this way fungal infection can be reduced. Transgenic potato plants with lysozyme gene providing resistance to Eswinia carotovora have been developed.

Delayed Fruit Ripening

The gas hormone, ethylene regulates the ripening of fruits, therefore, ripening can be slowed down by blocking or reducing ethylene production. This can be achieved by introducing ethylene forming gene(s) in a way that will suppress its own expression in the crop plant. Such fruits ripen very slowly (however, they can be ripen by ethylene application) and this helps in exporting the fruits to longer distances without spoilage due to longer-shelf life. The most common example is the 'Flavr Savr' transgenic tomatoes, which were commercialised in U.S.A in 1994. The main strategy used was the antisense RNA approach.

In the normal tomato plant, the PG gene (for the enzyme polygalacturonase) encodes a normal mRNA that produces the enzyme polygalacturonase which is involved in the fruit ripening. The complimentary DNA of PG encodes for antisense mRNA, which is complimentary to normal (sense) mRNA. The hybridisation between the sense and antisnse mRNAs renders the sense mRNA ineffective.

Consequently, polygalacturonase is not produced causing delay in the fruit ripening. Similarly strategies have been developed to block the ethylene biosynthesis thereby reducing the fruit ripening. *E.g.*, transgenic plants with antisense gene of ACC oxidase (an enzyme involved in the biosynthetic process of ethylene) have been developed.

In these plants, production of ethylene was reduced by about 97 per cent with a significant delay in the fruit ripening. The bacterial gene encoding ACC deaminase (an enzyme that acts on ACC and removes amino group) has been transferred and expressed in tomato plants which showed 90 per cent inhibition in the ethylene biosynthesis.

Male Sterility

The plants may inherit male sterility either from the nucleus or cytoplasm. It is possible to introduce male sterility through genetic manipulations while the female plants maintain fertility. In tobacco plants, these are created by introducing a gene coding for an enzyme (barnase, which is a RNA hydrolysing enzyme) that inhibits pollen formation.

This gene is expressed specifically in the tapetal cells of anther using tapetal specific promoter TA29 to restrict its activity only to the cells involved in pollen production. The restoration of male fertility is done by introducing another gene barstar that suppresses the activity of barnase at the onset of the breeding season. By using this approach, transgenic plants of tobacco, cauliflower, cotton, tomato, corn, lettuce, etc., with male sterility have been developed.

GENE TRANSFER METHODS IN PLANTS

To achieve genetic transformation in plants, we need the construction of a vector (genetic vehicle) which transports the genes of interest, flanked by the necessary controlling sequences, *i.e.*, promoter and terminator, and deliver the genes into the host plant. The two kinds of gene transfer methods in plants are:

Vector-mediated or Indirect Gene Transfer

Among the various vectors used in plant transformation, the Ti plasmid of *Agrobacterium tumefaciens* has been widely used. This bacteria is known as "natural genetic engineer" of plants because these bacteria have natural ability to transfer T-DNA of their plasmids into plant genome upon infection of cells at the wound site and cause an unorganised growth of a cell mass known as crown gall. Ti plasmids are used as gene vectors for delivering useful foreign genes into target plant cells and tissues. The foreign gene is cloned in the T-DNA region of Ti-plasmid in place of unwanted sequences.

To transform plants, leaf discs (in case of dicots) or embryogenic callus (in case of monocots) are collected and infected with *Agrobacterium* carrying recombinant disarmed Ti-plasmid vector. The infected tissue is then cultured (co-cultivation) on shoot regeneration medium for 2-3 days during which time the transfer of T-DNA along with foreign genes takes place.

After this, the transformed tissues (leaf discs/calli) are transferred onto selection cum plant regeneration medium supplemented with usually lethal concentration of an antibiotic to selectively eliminate non-transformed tissues. After 3-5 weeks, the regenerated shoots (from leaf discs) are transferred to root-inducing medium, and after another 3-4 weeks, complete plants are transferred to soil following the hardening (acclimatisation) of regenerated plants. The molecular techniques like PCR and southern hybridisation are used to detect the presence of foreign genes in the transgenic plants.

Vectorless or Direct Gene Transfer

In the direct gene transfer methods, the foreign gene of interest is delivered into the host plant cell without the help of a vector. The methods used for direct gene transfer in plants are: Chemical mediated gene transfer, *e.g.*, chemicals like polyethylene glycol (PEG) and dextran sulphate induce DNA uptake into plant protoplasts. Calcium phosphate is also used to transfer DNA into cultured cells. Microinjection where the DNA is directly injected into plant protoplasts or cells (specifically into the nucleus or cytoplasm) using fine tipped (0.5 - 1.0 micrometer diameter) glass needle or micropipette. This method of gene transfer is used to introduce DNA into large cells such as oocytes, eggs, and the cells of early embryo.

Electroporation involves a pulse of high voltage applied to protoplasts/ cells/ tissues to make transient (temporary) pores in the plasma membrane which facilitates the uptake of foreign DNA. The cells are placed in a solution containing DNA and subjected to electrical shocks to cause holes in the membranes. The foreign DNA fragments enter through the holes into the cytoplasm and then to nucleus. Particle gun/Particle bombardment - In this method, the foreign DNA containing the genes to be transferred is coated onto the surface of minute gold or tungsten particles (1-3 micrometers) and bombarded onto the target tissue or cells using a particle gun (also called as gene gun/shot gun/microprojectile gun). The microprojectile bombardment method was initially named as biolistics by its inventor Sanford (1988). Two types of plant tissue are commonly used for particle bombardment- Primary explants and the proliferating embryonic tissues.

Transformation - This method is used for introducing foreign DNA into bacterial cells, *e.g.* E. Coli. The transformation frequency (the fraction of cell population that can be transferred) is very good in this method. *E.g.*, the uptake of plasmid DNA by E. coli is carried out in ice cold CaCl2 (0-50C) followed by heat shock treatment at 37-450C for about 90 sec. The transformation efficiency refers to the number of transformants per microgram of added DNA. The CaCl2 breaks the cell wall at certain regions and binds the DNA to the cell surface.

Conjunction: It is a natural microbial recombination process and is used as a method for gene transfer. In conjunction, two live bacteria come together and the single stranded DNA is transferred via cytoplasmic bridges from the donor bacteria to the recipient bacteria. Liposome mediated gene transfer or Lipofection - Liposomes are circular lipid molecules with an aqueous interior that can carry nucleic acids. Liposomes encapsulate the DNA fragments and then adher to the cell membranes and fuse with them to transfer DNA fragments. Thus, the DNA enters the cell and then to the nucleus. Lipofection is a very efficient technique used to transfer genes in bacterial, animal and plant cells.

Selection of Transformed Cells from Untransformed Cells

The selection of transformed plant cells from untransformed cells is an important step in the plant genetic engineering. For this, a marker gene (*e.g.*, for antibiotic resistance) is introduced into the plant along with the transgene followed by the selection of an appropriate selection medium (containing the antibiotic). The segregation and stability of the transgene integration and expression in the subsequent generations can be studied by genetic and molecular analyses (Northern, Southern, Western blot, PCR).

Vector-mediated or Indirect Gene Transfer

Among the various vectors used in plant transformation, the Ti plasmid of *Agrobacterium tumefaciens* has been widely used. This bacteria is known as "natural genetic engineer" of plants because these bacteria have natural ability to transfer T-DNA of their plasmids into plant genome upon infection of cells at the wound site and cause an unorganised growth of a cell mass known as crown gall. Ti plasmids are used as gene vectors for delivering useful foreign genes into target plant cells and tissues.

The foreign gene is cloned in the T-DNA region of Ti-plasmid in place of unwanted sequences.To transform plants, leaf discs (in case of dicots) or embryogenic callus (in case of monocots) are collected and infected with *Agrobacterium* carrying recombinant disarmed Ti-plasmid vector. The infected tissue is then cultured (co-cultivation) on shoot regeneration medium for 2-3 days during which time the transfer of T-DNA along with foreign genes takes place. After this, the transformed tissues (leaf discs/calli) are transferred onto selection cum plant regeneration medium supplemented with usually lethal concentration of an antibiotic to selectively eliminate non-transformed tissues. After 3-5 weeks, the regenerated shoots (from leaf discs) are transferred to root-inducing medium, and after another 3-4 weeks, complete plants are transferred to soil following the hardening (acclimatisation) of regenerated plants. The molecular techniques like PCR and southern hybridisation are used to detect the presence of foreign genes in the transgenic plants.

GENE TRANSFER VIA GENETIC TRANSFORMATION

TRANSFORMATION, TRANSDUCTION AND TRANSFECTION –GENE TRANSFER METHODS

The three very effective modes of gene transfer Transformation, Transduction and Transfection observed in bacteria fascinated the scientist leading to the development of molecular cloning. The basic principle applied in molecular cloning is transfer of desired gene from donor to a selected recipient for various applications in the field of medicine, research, gene therapy with an ultimate aim of beneficial to the mankind.

- *Transformation:* Transformation is the naturally occurring process of gene transfer which involves absorption of the genetic material by a cell through cell membrane causing the fusion of the foreign DNA with the native DNA resulting in the genetic expression of the received DNA. Transformation is usually a natural method of gene transfer but as a result of technological advancement originated the artificial or induced transformation. Thus there are two types called as natural transformation and artificial or induced transformation. In natural transformation, the foreign DNA attaches itself to the host cell DNA receptor and with the help of the protein DNA translocase it enters the host cell. The presence of nucleases restricts the entry of two strands of the DNA, destroys a single strand thus allowing only one strand to enter the host cell. This single stranded DNA mingles with the host genetic material successfully.

  The artificial or induced method of transformation is done under laboratory condition which is either a chemical mediated gene transfer or done by electroporation. In the chemical mediated gene transfer, the cold conditioned cells in calcium chloride solution are exposed to sudden heat which increases the permeability of the cell membrane allowing the foreign DNA. The electroporation method as the name indicates, pores are made in the cell by exposing it to suitable electric field, allowing the entry of the DNA. The opened up portions of the cell are sealed by the ability of the cell to repair.

- *Transduction:* In transduction, a media like virus is required between two bacterial cells in transferring genes from one cell to the other. Researchers used virus as a tool to introduce foreign DNA from the selected species to the target organism. Transduction mode of gene transfer follows either a lysogenic phase or lytic phase. In the lysogenic phase, the viral (phage) DNA once joining the bacterial DNA through transduction stays dormant in the following generations. The induction of lysogenic cycle by an external factor like UV light results in lytic phase. In lytic phase, the viral or phage DNA exists as a separate entity in the host cell and the host cell replicates viral DNA mistaking it for its own DNA. As a result many phages are produced within the host cell and when the number exceeds it causes the lysis of the host cell and the phages exits and infects other cells. As this process involves existence of both the genome of the phage and the genome of the bacteria in the same cell, it may result in exchange of some genes between the two DNA. As a result, the newly developed phage leaving the cell may carry a bacterial gene and transfer it to the other cell it infects. Also some of the phage genes may be present in the host cell. There are two types of transduction called as generalised transduction in which any of the bacterial gene is transferred via the bacteriophage to the other bacteria and specialised transduction involves transfer of limited or selected set of genes.
- *Transfection:* One of the methods of gene transfer where the genetic material is deliberately introduced into the animal cell in view of studying various functions of proteins and the gene. This mode of gene transfer involves creation of pores on the cell membrane enabling the cell to receive the foreign genetic material. The significance of creating pores and introducing the DNA into the host mammalian cell contributed to different methods in transfection. Chemical mediated transfection involves use of either calcium phosphate or cationic polymers or liposomes. Electroporation, sonoporation, impalefection, optical transfection, hydro dynamic delivery are some of the non- chemical based gene transfer. Particle based transfection uses gene gun technique where a nanoparticle is used to transfer the DNA to host cell or by another method called as magnetofection. Nucleofection and use of heat shock are the other evolved methods for successful transfection.

Regeneration System

The development of an efficient system for genetic transformation is a valuable extension of the gene transfer tools for further crop improvement. In order to establish a successful *Allium* genetic transformation system, two key factors should be taken into account. One is the development of sophisticated methods to recover intact plants, either from fully dedifferentiated tissue or from organised tissues that are easy to regenerate. The other is the refinement of methods for the introduction of exogenous DNA into *Allium* germplasm.

In *Allium*, various plant regeneration systems have been developed using different starting material. Eady (1995) reviewed the different source materials used for *in vitro* culture of *Allium* species. The most successful regeneration systems in *Allium* use (im)mature embryos, root tips (segments), flower buds, suspension cultures or protoplasts as starting materials. A key point in *Allium* regeneration protocols is the fact that tissues are used consisting of actively dividing cells, such as mature or immature embryos of onion and shallot or calli induced from apical and non-apical root segments of *in vitro* plantlets of garlic. Such young and actively dividing callus material is uniquely suitable for genetic transformation.

Transformation System

Transformation of recalcitrant monocots like rice, wheat, barley and maize has been achieved by using direct gene transfer systems: chemical methods, electroporation, particle bombardment and silicon carbide fibres. Recently, *Agrobacterium*-mediated transformation of monocots has gained favour and many transgenic plants have been obtained using specific *Agrobacterium* strains. *Agrobacterium*-mediated transformation offers several advantages over other systems, the most important being the capability of delivering a single or low number of intact copies of relatively large segments of foreign DNA. Nowadays *Agrobacterium tumefaciens*-mediated transformation is routinely utilised in gene transfer to monocotyledonous plants, such as rice and maise (Hiei *et al.*, 1997).

Recently, reports were published showing that genetic transformation has become possible in *Allium* and this without any doubt is a major step forward (for review, Eady 2002). The progress of transformation research in *Allium* in the last 10-15 years. With report to particle bombardment, Klein *et al.* (1987) developed as the first one a high-velocity microprojectile method and demonstrated that epidermal tissue of onion could take up foreign DNA sequences. Wang (1996) obtained transgenic leek plants by particle bombardment with the *barnase* and *barstar* genes, and it was shown that the genes were present in the leek genome. Transient expression was shown with particle bombardment in garlic.

Park et al. (2002) and Sawahel (2002) reported that transgenic garlic plants were generated by particle bombardment. With respect to *Agrobacterium* mediated gene transfer, Dommisse *et al.* (1990) demonstrated that onion is also a host for *Agrobacterium* as evidenced by tumorigenic responses and opine production inside these tumours. Eady et al. (2000) developed a stable transformation protocol using immature embryos of *A. cepa* via *Agrobacterium tumefaciens*. Kondo *et al.* (2000) used highly regenerative calli derived from shoot primordial-like tissues to produce transgenic garlic plants by *Agrobacterium*-mediated gene transfer.

Zheng *et al.* (2001a, 2001b, 2004a, 2004b) developed a reproducible and stable transformation protocol using calli derived from mature embryos of onion and shallot or using calli derived from apical and non-apical root segments of *in vitro* plantlets of garlic via *Agrobacterium tumefaciens*. Due to the aforementioned efforts in developing reliable transformation systems for *Allium* crops nowadays transgenic shallot and garlic plants containing *Bt* resistance genes have been produced which confer resistance to beet armyworm (*Spodoptera exigua*) (Zheng *et al.* 2004a, 2004b). Furthermore, transgenic plants containing herbicide resistance and antisense versions of alliinase genes are available (Eady 2002).

SAFETY ASSESSMENT OF GENETICALLY MODIFIED (GM) ALLIUM CROPS

Genetic engineering of plants to produce genetically modified organisms (GMOs) was initiated nearly three decades ago with the intent to develop improved foods and medicines. In the year 2001, there were about 50 million hectares of GM crops grown worldwide (James 2001). Using GM technology transfer of genes into crops from any class of living organism has become a reality thereby producing novel kinds of crops.

Because of this, there is international agreement that a comprehensive safety assessment should be carried out before GM crop plants are grown commercially in agriculture. As transgenic herbicide and insecticide resistant *Alliums* have become a reality, it becomes important to analyse their safety aspects. We will try to do this using transgenic *Bt* shallots and garlic as an example.

To answer the political impact of GM crops a number of questions can be raised:

- How does the introduced gene changes the modified crop? Based on our experience, the phenotypic appearance of the crop due to the *Bt* transgene does not significantly change.
- Is there evidence of toxicity and/or allergenicity? This question has not been studied yet, but based on the experience with *Bt* corn and soybean it is not been expected.
- What are the effects on friendly organisms in the environment? This is not yet clear as field experiments have to be carried out. But large effects are not be expected based upon the experience gained from corn and soybean.
- Does *Bt Alliums* induce weediness in agricultural habitats or invasiveness in natural habitats? This does not seem to be a real problem as *Bt* shallots and garlic because these transgenics are only

modified into the direction of resistance to beet armyworm. As it is not known from literature that beet armyworm is involved in weediness or invasiveness, we do not expect major problems in this respect.

- Does gene flow between *Bt Alliums* and wild relatives exist and does it pose a threat? Gene flow between *Allium* species does exist, but it is very limited under greenhouse controlled condition (for review, Kik, 2002). Flowering of *Bt* shallots and garlic under current cultivation practice occurs sporadically. Therefore, the risk of gene exchange is very limited.

Insecticides based on endotoxin proteins of *Bacillus thuringiensis* have been in use for 40 years, and have a safety record for non-target invertebrates and vertebrates including mammals that far surpasses that of any synthetic chemical insecticide. This safety record, combined with the efficacy of certain *Bt* Cry proteins and the advent of recombinant DNA technology, led to the development of transgenic insect-protected Bt crops that are being adopted rapidly, especially *Bt* cotton and *Bt* maize, by farmers in the United States and a few other countries (Federici, 2002).

From the current state of knowledge, the impact of free DNA of transgenic origin is likely to be negligible (Dale *et al.* 2002). There is no reason at present to think that these crops present risks greater than those associated with the consumption of non-Bt crops. In fact, Bt crops may be safer for human consumption than conventional crops because they cobtain lower levels of mycotoxins and residues of chemical insecticides. All in all, GM safety issue in *Bt* garlic and shallots seem to be minimal. However only field and greenhouse experiments will give the crucial data needed to answer these aspects and make these transgenics acceptable for the general public.

DIRECT GENE TRANSFER METHODS

There are five different methods that transfer foreign DNA of desired function into plant cells. These methods are briefly discussed herewith.

Stimulation by Chemical Helpers

In 1977, Lurquin and Kado demonstrated the stimulation by poly-L-ornithine in uptake of plasmid DNA by cowpea protoplasts. In the presence of calcium ions, polyethylene glycol (PEG) also enhances DNA uptake by the protoplasts in the range 10"2 and 10"5.

Liposome-entrapped Plasmid Transfer

Liposomes are microscopic lipid molecules. These are produced when phospholipids are dispersed in aqueous phase. Charges can be developed on liposomes as desired. Liposomes entrap DNA and interact with many cells in such a way that plasmid will be transferred into plant genome.

For example, unilamellar liposomes have been used for transformation and expression of a cloned cDNA copy of potato spindle tuber viroid (PSTV) and a chimeric DNA of E. coli plasmid into protoplasts.

Electroporation

Introduction of foreign DNA into the living cells (as protoplasts) by exposure of brief electric pulse (250-350 V) is known as electroporation. Minimum manipulation of protoplasts ensures their high survival rates. Using electroporation, transient expression of foreign DNA has been demonstrated in maize, wheat and rice.

Microinjection

In 1986, Crossway and co-workers used microinjection technique for direct gene transfer into plant cells. Animal cells can be transformed at high frequency by microinjection than the plant cells. In this technique protoplasts are surface attached on a slide by embedding in agarose using a holding pipette. DNA is transferred by injecting a microneedle or a glass micropipette having fine tip CO.5 - 1μ diameter). The microinjected protoplasts are cultured as described earlier.

Particle (Microprojectile) Bombardment Gun

It is also called as particle gun, shot gun, microprojectile gun and biolistic missile. In this method foreign DNA containing desired genes are coated onto the surface of tungusten or gold particles (4 micrometer diameter).

These particles are bombarded to a high velocity (by particle gun) onto plant cells or tissues. The particles pierce cell walls and membrane.

VECTOR-MEDIATED GENE TRANSFER IN PLANT CELLS

Agrobacterium consists of Ti-plasmid which is directly responsible for tumour induction. Transfer of small DNA segment (T-DNA) from Ti-plasmid and its integration into host genome induces tumour formation in plants. T-DNA region of Ti-plasmid is essential for Agrobacterium-mediated transformation of plants. The Ti-plasmid is taken out from A. tumifaciens and a foreign DNA of desired function is inserted into T- DNA region by using restriction enzyme. The recombinant T-plasmid is incorporated into A. tumifaciens cells. The steps of Agrobacterium-mediated gene transfer in a plant.

Co-cultivation with Protoplasts

Freshly isolated protoplasts are co-cultivated with transformed A. tumifaciens for a few days. Then the protoplasts are washed and cultured in medium containing antibiotic. The regenerated protoplasts are transferred onto culture medium that enables the development of transformed callus and plants.

Leaf-disc Infection Method

In this approach leaves (of tomato, tobacco, Petunia, etc.) are surface sterilised and inoculated with a genetically modified A. tumifaciens cells. Then the leaves are cultured for two days. During this time infection takes place. The leaf discs containing infection are transferred to a selection cum regeneration medium containing kanamycin. Kanamycin selectively eliminates the non-transformed cell/tissues. Within 2-4 weeks, these cultures regenerate into transformed plantlets.

Then, the genetically modified plantlets are transferred into soil after hardening of regenerated plant so that they can be acclimatised with the new environment. Presence and function of foreign gene in transgenic plant are found out by molecular techniques as given in transgene analysis.

TRANSGENE ANALYSIS

In plant transformation experiments selection of transformed cells from the mixed populations of both transformed and untransformed cells is an important event. A marker gene (*e.g.,* antibiotic resistance or herbicide resistance) is also introduced along with foreign DNA into the host cells. A selection medium is prepared into which antibiotics or herbicide (for which marker gene has been introduced) is mixed. The transformed cells/tissues are cultured on selection medium.

Only those cells will grow on such medium which contains marker gene such cells are said to be transformed. The untransformed cells will not grow because they lack marker gene. After selection subsequent maintenance of marker in transgenic plant is not necessary. Since the herbicide-resistance gene may be transformed to bacterium or herbicide-resistance gene may be transferred to weeds, the marker genes must be eliminated. The removal of marker enables the transgene to function and express properly. Integration and expression of transgenes are further studied following molecular techniques like PCR, Southern blot, Northern blot, Western blot, etc.

7

Stem Cell Bioengineering and Tissue Regeneration

INTRODUCTION

At the interface of micro- and nanofabrication and experimental biology lies enormous potential to address important problems in biology and medicine. This is because biological systems are highly complex and cannot be easily understood without tools that can study such complexity at length scales that are relevant to biological systems. This complexity extends from developmental processes, where a single cell undergoes many rounds of division and proliferation to become a fully developed organism, to the maintenance and regeneration of adult tissues.

Cellular processes are controlled by the genetic factors in each cell, which are not only intrinsically regulated but are also controlled by the local cellular microenvironment. Thus, the local chemical, biological, and mechanical environments can provide a coordinate set of regulatory cues to control cell behavior. For example, in the early stages of development, embryonic cells in the inner cell mass communicate with one another through paracrine and autocrine signaling and cell—cell contacts that influences organogenesis. Cellular communication and signaling occurs through a number of different mechanisms, including direct cell—cell contact, soluble factors, and cell—matrix interactions. Soluble factors and signaling molecules provide different cues depending on the identity, concentration, and context. Furthermore, mechanical forces imparted by the surrounding extracellular matrix (ECM) can signal cells to specific fate decisions.

The complexity of the interactions between cells and the microenvironment is not only found during development and organogenesis but also found in mature organisms. For example, blood cells are continually replenished throughout an organism's lifetime. It is known that hematopoietic stem cells, which reside in the bone marrow, interact closely with the surrounding endothelial cells, osteoblasts, and fibroblasts that regulates their self-renewal and differentiation toward different blood cell fates.

Our research group, along with others, is interested in developing technologies at the interface between micro- and nanoengineering and materials science for engineering controlled microenvironments that can aid in understanding the interactions between cells and their microenvironment. The goal is to overcome a major experimental challenge in experimental biology, which is to recreate the in vivo cell— microenvironment interactions in vitro. In a tissue culture Petri dish, cells interact with a two-dimensional plastic surface that is drastically different from the environment that is found in vivo.

A salient example of the deleterious effects of disrupting the natural cellular microenvironment is the loss of liver hepatocyte function on culture outside the body. In the liver, hepatocytes reside in controlled tissue units called liver lobules, which have a high degree of control and complexity. Liver lobules are made from organized hepatocytes that are assembled in hexagonal structures that are highly vascularized and are in intimate contact with the surrounding endothelial cells. The inability to recreate this complex microarchitecture in vitro may be a key reason that the significant regeneration capacity of the liver and its diverse metabolic functions have not been recreated in vitro.

Our research group aims to develop and use technologies to recreate cell—microenvironment interactions to produce in vitro culture conditions that can be used for understanding cell biology or to generate three-dimensional (3D) tissue constructs for cell-based therapies. Furthermore, engineered tissues can be used for screening drugs and to investigate the underlying mechanisms of disease. These modulations of the cellular microenvironment will be made by using microfabrication to sculpt and assemble advanced materials to control the interactions of cells with the microenvironment. For example, through the use of microfluidics, it is possible to control the temporal and contextual presentation of soluble factors to induce cellular events; patterned surfaces can be used to generate spatially controllable cocultures to control cell—cell interactions; and, cell-laden hydrogels can be created to investigate cell behavior in 3D. Finally, it is possible to perturb mechanical properties of materials and local sheer stresses induced by fluids to modify cell responses.

During the past several decades many tools required to investigate biology at the micro- and nanometer length scales have been developed. There is still much work to be accomplished, but great strides have been made in making micro- and nanofabricated systems much more accessible to common laboratory use. For example, with rapid prototyping or soft lithography, micro- and nanopatterned silicon wafers, or other templates, can be quickly and easily replicated with an elastomeric copolymer such as polydimethylsiloxane (PDMS). PDMS replicas can be used as microfluidic channels, for molding of biomaterials or as stamps to pattern surfaces. The power of these techniques is that they can be used to control the architecture of materials at length scales much smaller (<100 nm) or much larger (>1 mm) than a typical cell.

In this article, we describe our laboratory's work in merging biomaterials and advanced fabrication techniques to control the cell microenvironment and engineer tissues with controlled microarchitecture. The common element of the various projects is that they use biomaterial microstructures to engineer cell aggregates and generate tissue constructs. We describe two different ways to engineer tissue-like assemblies with controlled microarchitectures by using either a top-down or bottom-up approach.

With a bottom-up or "Lego-like" approach, we create small microfabricated tissue units that assemble into larger tissuelike constructs with controlled microarchitectures. With a top-down engineering approach, we sculpt biomaterials into micro- and nanoscale structures that mimic tissue constructs. In each of our approaches, we use microfabrication techniques to mold and control the size, shape, and microscale features of biocompatible hydrogels. We use photo-and soft lithography as well as micromolding techniques as they are simple methods that can be easily made compatible with a variety of different materials.

STEM CELL BIOENGINEERING

Stem cells have generated more excitement, scrutiny and controversy than any other area of recent scientific study. The first stem cells, which were discovered half a century ago, were isolated from blood. Now, scientists around the globe are researching various types of stem cells for their potential to regenerate lost tissue and revolutionize medicine.

Embryonic stem (ES) cells are derived from the embryo when it exists as a blastocyst. They have the ability to develop into all the different cell types found in the body. Actually, when a sperm fertilizes an egg, the resulting single cell begins to divide and multiply at a rate much faster than that observed in somatic cells. Scientists refer to these cells as totipotent stem cells. These primordial embryonic cells have the potential to grow into a complete mammal. Within days of fertilization, these new and dividing cells form a hollow sphere, called a blastocyst. Stem cells arising in the inner mass of the blastocyst are called the ES cells.

The ES cells are considered pluripotent – they can divide indefinitely and blossom into all the various tissue types of the human body, but they have the lost the totipotent ability to grow into a separate being. After roughly 14 days, ES cells divide to give rise to what will eventually develop into the spine. At this stage, the stem cells within the embryo are considered multipotent. These stem cells can grow into some tissues, but not all tissues. Those destined to become bone or blood, for example, may not be able to form stomach or skin.

By definition, stem cells have two important characteristics that distinguish them from other types of cells. First, they are unspecialized cells that renew themselves for long periods of time through cell division of at least one daughter cell. Secondly, under certain physiological or experimental conditions, they can be induced to differentiate. This means that they can divide into cells with special functions, such as the beating cells of the heart muscle or the insulin-producing cells of the pancreas.

Discovery of ES Cells

The work that laid the foundation for ES cell discovery was the study of teratocarcinomas, complex tumors containing a mix of specialized cell types as well as a population of unspecialized cells. These unspecialized cells are called embryonal carcinoma (EC) cells. The latter were shown to be pluripotent and could give rise to various cell types both *in vitro* and *in vivo*. It was therefore natural to consider using these cells for therapeutic purposes.

However, EC cells never seemed ideal for this purpose because they had an abnormal number of chromosomes and originated from tumors. Careful study of the induction of teratocarcinomas in experimental animals, as well as an understanding of the biology of EC cells and early embryos, led scientists to the discovery of ES cells in the early 1980s. The demonstration that ES cells contained the normal number of chromosomes and were truly pluripotential has influenced many scientific disciplines.

Biological Role and Properties of Stem Cells

Stem cells differ from other kinds of cells in the body. All stem cells, regardless of their source, have three general properties: they are capable of dividing and renewing themselves for long periods; they are unspecialized; they can choose to become one of the many different types of cells present in the body based on signals from their environments.

Finding the answers to two fundamental questions about stem cells that relate to their long-term self-renewal is crucial to our ability to successfully grow these cells in laboratory and in turn use them for various tissue engineering and cellular therapies. The first question deals with why embryonic stem cells can proliferate for extended periods of time in the laboratory without adopting a specialized fate, while most adult stem cells cannot.

The second issue addresses which factors in living organisms normally regulate stem cell self-renewal and differentiation. Discovering the answers to these questions may make it possible to understand how cell proliferation is regulated during normal embryonic development or during the abnormal cell division that ultimately leads to cancer.

Stem Cells are Unspecialized

One of the fundamental properties of stem cells is that they do not perform specialized functions. A stem cell cannot pump blood through the body (like a heart muscle cell) or supply oxygen to the cells of the body (like a red blood cell). However, unspecialized stem cells can give rise to specialized cells including heart muscle cells, blood cells, or nerve cells. They do this by coordinating their gene expression in an elaborate and complex pattern spanning many generations of cells.

Stem Cells are Self- renewing

Most specialized cells in our body (such as muscle cells, blood cells, or nerve cells) do not replicate themselves. The supply of these cells is maintained by stem cells, which replicate many times and then differentiate into the specialized cells that are needed. In this way, cells are continuously replenished as they die. This is called homeostasis.

When cells replicate themselves many times over it is called proliferation. A starting population of stem cells that proliferates for many months in the laboratory can yield hundreds of millions of cells. If the resulting cells are identical (in terms of being unspecialized themselves and capable of generating specialized cells) to the parent stem cells, the ES cells are said to be capable of long-term self-renewal. Scientists are greatly interested in determining specific factors and conditions that allow stem cells to remain unspecialized in the laboratory for long periods of time.

Stem Cells can give Rise to Specialized Cells

Differentiation is the process whereby stem cells give rise to specialized cells. Scientists are just beginning to understand the signals that trigger stem cell differentiation. These signals include chemicals secreted by other cells, physical contact with neighboring cells, and contact with molecules in the microenvironment.

Can specific sets of signals be identified as responsible for the promotion of differentiation into specific cell types? Answering this question is critical in order to learn how to maintain pools of stem cells. The development of stem cell-based technologies will depend on our ability to reproducibly drive the differentiation of stem cells into specific tissue lineages. This issue is of particular importance for the development of ES cell-based technologies, where the differentiation of certain subsets of cells and tissues is uncommon. The generation of ES cell-derived heart muscle cells is one example where significant improvements in tissue-specific differentiation are needed. Under default mechanisms, only approximately 1 per cent of ES cells form cardiac tissue.

Growing Embryonic Stem Cells in the Laboratory

The process of growing cells in the laboratory is known as cell culture. Mouse (or human) embryonic stem cells are isolated by transferring the inner cell mass of the blastocyst into a plastic laboratory culture dish containing medium, which provides nutrients and necessary growth factors. The cells divide and spread over the surface of the dish. In the case of human ES cells, the inner surface of the culture dish is typically coated with mouse embryonic skin cells, which are non-dividing and secrete a host of growth factors. This is known as a feeder layer and is necessary for the prevention of ES cell differentiation. The feeder layer provides a sticky surface to which the inner cell mass cells can attach. Mouse ES cells, however, can be maintained in culture without the need of a feeder layer. In this case, the culture medium is supplemented with high concentrations of a protein known as leukemia inhibitory factor (LIF), which prevents differentiation.

Over the next several days, the cells of the inner cell mass grow and divide and spread on the culture dish. When the cell concentration is high enough, they are removed gently and plated into several fresh culture dishes. This process of replating the cells is repeated for many months, and is called subculturing. After six months or more, the original 20 to 30 cells of the inner cell mass yield many millions of embryonic stem cells. When embryonic stem cells have proliferated in cell culture for six or more months without differentiating, are pluripotent, and appear to be genetically normal, they are referred to as an ES cell line. Once ES cell lines are established, batches of them can be frozen and kept for further experiments or even shared with other laboratories for further culture and experimentation.

ES Cell Differentiation

As long as mouse ES cells in culture are grown in the presence of LIF or a feeder layer, they can remain undifferentiated (unspecialized). If LIF is removed from the culture medium, cells will begin to clump together to form 3-dimensional structures called embryoid bodies (EBs). The cells then begin to differentiate spontaneously. Allowing spontaneous differentiation is a good way to determine if a culture of embryonic stem cells is healthy. It is not an efficient way to produce cultures of a specific cell type. In order to generate a large number of cells of a specific type, scientists use the process of directed differentiation. This is done by changing the chemical composition of the culture medium, altering the surface of the culture dish, or modifying the cells by inserting specific genes.

Adult Stem Cells

Many adult tissues (such as the bone marrow, brain and gut) contain stem cells. Like ES cells, adult stem cells can make identical copies of themselves for long periods of time (self-renewal). At the same time, they can give rise to mature cell types that have characteristic shapes and specialized functions. Stem cells typically generate an intermediate cell type(s) before they achieve their fully differentiated state. The intermediate cell is called a progenitor cell. Progenitor cells are partly differentiated cells in the sense that they are committed to a particular cell lineage and, upon division, and give rise to differentiated cells. Of all the adult stem cells identified thus far, Hematopoeitic stem cells (HSCs) are the best characterized.

Stem Cell Plasticity

Until recently, adult stem cells were considered to be irreversibly committed to specific lineages of differentiation. HSCs, for example, normally give rise to all types of blood cells such as red blood cells, white blood cells and platelets. It was previously believed that HSCs could not give rise to any cells of a different tissue. However, a number of experiments over the last several years have raised the possibility that stem cells from one tissue may be able to give rise to cell types of a completely different tissue. This phenomenon is known as 'stem cell plasticity'.

Examples of such plasticity include bone marrow stem cells becoming neurons, or pancreatic islet cells that are capable of producing insulin. Exploring the possibility of using adult stem cells for cellular therapies has recently become an active area of research. The apparent plasticity of adult stem cells has forced scientists to reconsider many fundamental concepts of stem cell biology. Many new questions arise from these studies. For instance, are the recently characterized stem cells typical tissue-specific stem cells? Might they be a sub-population of stem cells with a developmental potential closer to that of embryonic stem cells? There is a possibility that stem cell plasticity results from in vitro manipulations and does not reflect normal behavior *in vivo*. Regardless, this phenomenon has important clinical implications.

Potential Therapeutic Applications of Stem Cells

The ultimate objective of stem cell bioengineering is to be able to understand and possibly control stem cell differentiation and lineage commitment in vitro. If this can be achieved, a multitude of therapeutic applications can be envisioned. One potential application is the generation of different types of neurons for the treatment of Alzheimer's disease, spinal cord injuries, or Parkinson's disease. The production of heart muscle cells for heart attack survivors may also be possible. The generation of insulin-secreting pancreatic islet cells for the treatment of type-1 diabetes, and even the generation of hair follicle stem cells for the treatment of certain types of baldness, have been considered.

Stem cells could also be useful for a number of tissue engineering applications such as the production of complete organs including livers, kidneys, eyes, hearts, or even parts of the brain. This represents a considerably greater challenge, beyond the generation of specialized cell types, and will require considerable time and effort to develop. Other areas that would benefit from a better understanding and control of stem cell proliferation in vitro are drug testing, cancer research, and fundamental research on embryonic development.

Human ES cells could potentially be used for all of the above applications. Human ES cells are obtained from aborted fetuses or fertilized eggs. This has come under ethical scrutiny since use of these procedures requires serious moral consideration by society. A possible way to circumvent this issue would be to use stem cells isolated from adult tissues. If, as recent studies have suggested, the plasticity of stem cells is greater than previously imagined, then it may be feasible to use adult stem cells as an initial cell population for therapeutic modality. Bone marrow, peripheral blood or umbilical cord blood HSCs could be ideal for such procedures. They are relatively easy to harvest and robust assays exist to quantify HSCs and distinguish them from more differentiated progenitors. Although much remains to be learned about the factors controlling their self-renewal and differentiation, the state of knowledge about HSCs is considerably more advanced than for other stem cell systems.

HYDROGEL MICROSTRUCTURES FOR STEM CELL BIOENGINEERING

Polymers such as poly(ethylene glycol) (PEG) are so hydrophilic that water complexes with their polymer chains to prevent subsequent protein adsorption or cell adhesion. We have been interested in exploiting this phenomenon to create microscale structures with controllable surface properties. A number of different microscale hydrogel structures made by micromolding of photocrosslinkable PEG. The engineered microwells can be made with, and without exposed underlying glass substrates that can regulate protein adsorption on the surfaces. As cells can adhere to adsorbed protein layers, it is possible to use micropatterning of PEG structures to immobilize cells inside PEG microwells.

It is important to note that fluid elements around micro-scale structures behave differently from those in macroscale systems. For example, mathematical modeling of arrays of microwells under fluid flow reveals that as the depth of the microwells increase or the diameter of the wells decrease, the regions in the wells become protected from shear stresses. Thus, cells in deeper- or-smaller diameter microwells that are exposed to less shear stress, are not washed away under flow, and can be stably integrated into the device. This system of arrayed microwells made from PEG hydrogel can be used to capture cells and other particles of interest in a rapid and controllable manner.

The flow patterns inside the structures can also be used to control cell location. For example, we have used mathematical simulations to show that when fluid flows over arrays of microwells that are less than 50 ìm in diameter, fluid recirculated in the structures, whereas in larger structures, fluid penetrated the structures. These results were also confirmed experimentally. In the 75-ìm grooves, there was no recirculating flow or back flow, and cells were somewhat evenly distributed across the width of the grooves.

However, in the 50-ìm grooves, recirculating flows aligned cells along the upstream side of the grooves. Arrayed microwells can be used as a platform to ask questions in stem cell biology and to direct cell fates for therapeutic applications. Specifically, the microwell platform is well suited to control the size of stem cell aggregates. In these aggregates, embryonic stem (ES) cells begin to initiate differentiation pathways that more closely mimic the developing embryo.

In a typical ES cell aggregate experiment, ES cells are plated in a nonadhesive dish, and cells cluster together into aggregates of varying sizes and begin to differentiate. PEG microwells can be used to control this process and form cell aggregates in a simple manner. Cells seeded inside the microwells are in low shear stress regions and remain in the microwells under flow. In this way, we can control the formation and size of cell aggregates and produce homogeneous populations of microsphere cell aggregates. The examples of controlled cell—cell interactions and the culturing of populations of cell aggregates, embryoid bodies (EBs), with homogeneous sizes. The EBs of varying sizes cultured in microwells and EBs after retrieval from the microwells after 10 days of culture.

The number of cells that form an EB can be used to direct stem cell differentiation. If one can control the number of cells per EB, then comparisons of small cell number and large cell number EBs are possible. In different sized EBs, individual cells will experience a different microenvironment and will be directed toward different fates.The outcomes of experiments with EBs of different sizes that were investigated with respect to the size-dependent expression of cardiac and endothelial cell differentiation. In this example, we examined the differentiation of EBs of three different sizes by using two different experimental protocols. In both cases, ES cells were cultured in microwells of different sizes for 5 days. In one protocol, we continued culturing the EBs in microwells for an additional 10 days.

Alternatively, we retrieved the EBs and replated them on Matrigel for an additional 10 days. In both cases, we analyze the outcomes with respect to gene expression and protein markers of cardiac and endothelial cell differentiation. In particular, we found that a higher fraction of larger EBs were spontaneously beating, a characteristic of differentiation into cardiac cells. In addition, the cardiac marker sarcomeric alpha acti-nin (SαA) was highly expressed in the beating cells. Large EBs (450 μm in diameter) cultured in the microwells showed strong SαA staining and beat spontaneously. In the smaller EBs (150 μm in diameter) that were cultured throughout the experiment in microwells, we saw minimal expression of SαM and a significant reduction in spontaneous beating. Larger EBs also had more cells with typical cardiac cell morphologies, when replated after 5 days of microwell culturing and when culturing in microwells for the course of the experiment.

Analysis of vascular differentiation showed an opposite response with smaller EBs giving rise to more endothelial-like cells. In these experiments, a higher degree of expression of the endothelial cell marker CD31 was observed in smaller EBs compared with larger EBs. Also, $CD31^+$ cells formed capillary-like structures. Additionally, smaller EBs that were replated on Matrigel started to form sprouts that stained positive for CD31 at a higher frequency and length than those observed in larger EBs, further suggesting that there was more endothelial cell differentiation in smaller EBs than in larger ones.

The microfabricated well arrays can also be used to generate homogeneous cultures for studying the molecular and biochemical mechanisms that regulate the observed size-dependent outcomes. In further analysis of the EBs of different sizes, we did not see significant differences in expression of a number of ECM components including fibronectin, collagen IV, and laminin a and b. We also analyzed the expression of Wnts, a family of signaling molecules that control a number of events during embryonic development. Although we did not observe differences in the expression of canonical Wnt family members, beta-catenin or Wnt2, we observed differences in expression of two member of the non-canonical Wnt family molecules.

In particular, it was observed that Wnt5a was highly expressed in smaller EBs, whereas Wnt11 was more prominently expressed in larger EBs. The size-dependent Wnt5a expression and the effect on cardiac and vascular differentiation were confirmed by siRNA silencing and also by exogenous addition of Wnt5a. In small aggregates with Wnt5a silencing, endothelial cell differentiation is significantly reduced. Addition of Wnt5a molecules to larger aggregates, which usually promote cardiac over endothelial differentiation, results in increased expression of endothelial markers. Additionally, the frequency of endothelial sprouts is increased. Taken together, these results show that differential expression of non-canonical Wnt molecules is highly important in EB size-dependent behavior of endothelial and cardiac cell differentiation.

These size-controlled EB experiments demonstrate the utility of microwell-based cell culture in answering interesting questions in stem cell biology; however, there are still many interesting aspects of the microwell culturing that have yet to be explored. For example, we have yet to investigate the effects of size restriction on stem cell proliferation and differentiation. Furthermore, the mechanisms that direct stem cell differentiation in these systems are yet to be fully elucidated.

Microwell technologies have many advantages over tradition cell culturing methods, such as the ability to control aggregate size, and the ability to protect cells from external shear stress. It is also possible to control microenvironmental factors including cell—cell and cell—matrix interactions with engineered hydrogel microwells. However, there are some potential shortcomings including an inability to individually address single microwells and an inability to dynamically control aggregate size as aggregates expand with cell proliferation. Additionally, aggregate size is controlled by restricting lateral growth, and during long-term culturing cell aggregates can proliferate outside of the bounds of the microwell structures. Retrieval of aggregates from the microwell structures has also proven challenging, and great care is needed to retrieve high yields of intact aggregates. With new microscale technologies and hydrogel microstructures, we are beginning to address some of these shortcomings.

BOTTOM-UP ASSEMBLY OF MICROGELS FOR TISSUE ENGINEERING

Natural tissues are made from highly complex tissue microarchitectures. A major challenge in tissue engineering is to recreate tissue constructs with appropriate microarchitectures and functions.

To this end, we have explored two different microengineering approaches:

1. The bottom-up assembly of microgels and
2. The top-down engineering of microscale biomaterials.

The former, bottom-up assembly, is a bioinspired approach to the assembly of complex tissue microarchitectures using microgel building blocks. In nature, many tissues are made up of assemblies of small tissue modules (i.e., repeating functional units). For example, muscles are made from bundles of myofibers; the liver is composed of lobules; the kidney is made up of nephrons; and the pancreas contains islets. These functional tissue units are well vascularized and are made of patterned assemblies of the units. In our laboratory, we are using a bottom-up approach to tissue engineering to mimic this concept. This is done by using microfabrication techniques to engineer microscale tissue units and then assembling the resulting structures to generate tissue-like complexity.

Hydrogels are attractive materials for making microengineered tissue building blocks because of their hydrated nature and biomimetic mechanical properties. We have previously used a number of methods to engineer cell-laden hydrogels of controlled shapes and sizes. For example, photocrosslinkable hyaluronic acid-based hydrogels were micro-molded into controlled shapes. To overcome the potential limitation of micromolding, which include its batch process, we have used microfluidic systems to create a continuous process to generate controlled shape microgels. In this process, the hydrogel precursor solution containing cells was flowed continuously past a light source to produce individual cell-laden microgels. A shutter was used to control the exposure of the hydrogel precursor to light, and a mask was used to control the microgel shape. In this way, we and our collaborators produced microgel building blocks in a continuous process.

We have also developed methods of inducing the assembly of microgels into tissue-like structures. The self-assembly of microgels differs substantially from molecular self-assembly in that the forces at work in molecular assembly differ from those that drive the assembly of microscale hydrogels. Thus, a key aspect that must be considered for microgel assembly is the type of forces that will be used to drive the assembly process.

One of our first ideas was to direct the assembly process by exploiting the tendency of hydrophobic and hydrophilic substances to minimize the interaction with each other. Specifically, we used the difference in chemical properties at a hydrophobic—hydrophilic liquid interface and at an air—liquid interface to drive the assembly process. Following on the work of others, we have used the forces at these interfaces to drive the assembly of microgel building blocks as the energy of the system is minimized.

Two different examples of self-assembled microgel structures. In one case, two differently shaped microgel building blocks are assembled using an oil—water interface. In the simple case of oil in water, the solutions minimize the surface interactions and droplets of water form in oil (or oil droplets in water). With microscale hydrogels in a bulk oil phase, a similar process occurs: the hydrogels are driven together as the surface interactions are minimized. Controlling the microgel shape leads to control of the assembly process. With appropriately shaped microgels, it is possible to create "lock-and-key" assemblies, where the shape of different microgels is matched to create mesoscale hydrogels with controlled microarchitectures. By using an oil—water interface, but we have also been able to create similar structures using an air—liquid interface. We have also been able to use this assembly process to create larger structures. By using two interfaces, oil—water and a wetting substrate, we can "wrap" assembled microgels around a wetting substrate to generate 3D structures.

The examples of bottom-up assembly of microgels demonstrate the potential utility of this technique to create complex tissue-like structures; however, there are some potential disadvantages with the process. For example, it is challenging to create structures that are 10s of centimeters in length. Such structures will require strong secondary crosslinking to ensure adequate bonding between microgels, and, if free standing, will require building blocks with mechanical properties sufficient to support the weight of the structure. Additionally, it would be advantageous to develop two-phase liquid systems other than mineral oil and water to drive the assembly process.

Further research is required to overcome these limitations, and we are actively pursuing microscale techniques to advance this bottom-up approach to tissue engineering. We are interested in exploring different aspects of the approach including the ordered and sequential assembly of microgels. We are also interested in exploring new hydrogels for microgel fabrication and developing composite hydrogels for controlling cell—ECM interactions within each microgel building blocks.

TOP-DOWN MICROSCALE TISSUE ENGINEERING

Some of the initial studies in the use of microfabrication techniques for tissue engineering were done in attempts to engineer microfluidic networks to recreate vascularized tissue structures in scaffolds made from materials such as silicon, PDMS, or degradable synthetic polymers. Although these materials are well suited for microscale fabrication, they are not amenable to cell encapsulation. Recently, we have been creating microfluidic channels with hydrogels made from natural polymers to enable the formation of hydrated vascular networks within which cells can be encapsulated in the bulk phase of the materials.

A micromolding approach to produce cell-laden hydrogel microfluidic channels. In this example, hepatocytes were embedded in agarose and molded into a microchannel. Perfusion within the hydrogel could be achieved through diffusion from the channels into the hydrogel walls and was confirmed in cross-sections of agarose microfluidic channels after only a few minutes of flow. In this way, the microfluidic hydrogel channels have the potential to mimic native vasculature, as oxygen and nutrients from liquid in the channels diffuse into the surrounding hydrogel.

This effect can be seen in cellladen agarose hydrogels with engineered microporosity. After 3 days of culture, a ring of viable cells can be seen on around the channel that corresponds to the diffusion pattern of oxygen and nutrients. With these types of micro-engineered structures, one can begin to create tissue-like constructs with encapsulated cells and endothelialized channels. Multiple constructs can be stack to create more complex vasculatures and tissue-like constructs.

The size, structure, and complexity of multilayered constructs are limited by the material properties of the hydrogels used to create the constructs. Although the examples presented here make significant progress in mimicking native vasculature, there are many technological challenges that remain. For example, microscale technologies that can create high densities of microscale channels with complex branching are required. Fabrication of microchannels that extend out in 3D and that have hierarchical structures are also needed. We are actively pursuing new microscale technologies to address these limitations.

ENGINEERING MICROENVIRONMENTS TO CONTROL STEM CELL FATE AND FUNCTION

Stem cells – defined by the capacities for self-renewal and for differentiation into multiple cell types – hold great promise for applications in regenerative medicine and as model systems for basic developmental biology studies. Tissue-specific adult stem cells have been isolated from a broad variety of adult tissues in many species. In addition to these multipotent stem cells, pluripotent embryonic stem cells that have the ability to generate all cells in an adult organism have been isolated from embryos, and induced pluripotent stem cells (iPS) have been created from somatic cells via the forced overexpression of key factors. Understanding how to preciously control the behavior of these cells is a key challenge for utilizing them in both therapeutic and scientific applications.

The hallmark properties of stem cell self-renewal and differentiation are governed by a complex set of extrinsic cues in collaboration with intrinsic gene regulatory machinery. The identification of numerous tissue-specific stem cells has driven recent investigations into whether there are intrinsic regulatory mechanisms shared by all self-renewing cells.

However, gene expression profiling studies across many different types of stem cells have not been able to define a common, core set of genes responsible for "stemness". This finding in part has led to a perceptible shift in focus from the role of intrinsic regulation to the function of cell extrinsic stimuli as determinants of cell fate. The conceptual framework for the latter efforts was initially proposed in 1978 by Schofield, who hypothesized that the soluble growth factors, extracellular matrix interactions, and cell-cell interactions that comprise the stem cell microenvironment play a role in supporting stem cell self-renewal. Accordingly, the identification of extrinsic factors and their downstream, intracellular signaling pathway targets has led to much progress in understanding how to control the expansion and directed differentiation of stem cells.

In parallel to these basic efforts, engineered microenvironments have been increasingly successful in controlling stem cell fate by emulating the key regulatory signals from native stem cell niches. This chapter has the dual objectives of discussing principles of stem cell biology, focusing on the strategies and mechanisms by which niche components support stem cell functions, as well as illustrating key examples of engineered stem cell microenvironments that build from these principles. We present examples of how growth factors and morphogens, neighboring cells, and the extracellular matrix regulate the size and composition of the niche.

Since there are many more cell types in organisms than there are signaling factors, biology encodes information that specifies cell fate not only through the presence or absence of signals, but also via their combinations, localization, level, and timing. The examples of engineered microenvironments discussed below indicate how a detailed understanding of how these components regulate stem cells in native niches has led to the elucidation and development of key design criteria for synthetic niches that allow greater control over cell fate.

GROWTH FACTORS AND MORPHOGENS

Immobilization of Signaling Factors as a Mechanism to Control Local Concentration

Growth factors and morphogens can exert potent, long-range effects in stem cell microenvironments. Owing to their relative ease of study, soluble growth factors and their downstream signal transduction pathways are the best characterized determinants of stem cell fate and have been extensively used in *ex vivo* stem cell culture systems, as has been effectively discussed elsewhere. We will instead discuss the application of growth factor immobilization, a ubiquitous theme in developmental biology, to engineered niches.

In vivo, numerous growth factors and morphogens are immobilized by binding to the extracellular matrix through specific heparin-binding domains or by direct binding to ECM molecules such as collagen as well as by fibronectin, or direct anchoring to cell membranes. Immobilization of growth factors in this manner can serve to increase local concentration of the protein by hindering diffusion and receptor-mediated endocytosis.

For example, the morphogen Sonic hedgehog is modified at its termini by lipids – cholesterol and palmitic acid – that link it to the cell membrane and thereby limit its mobility. Removing the lipids dilutes the factor to a lower concentration and thereby shrinks the range of its effective activity. Accordingly, mimicking the natural immobilization of cytokines is one approach utilized by engineers to concentrate factors in proximity to the cell surface in a manner that effectively activates target signaling pathways, as well as reduces the levels of growth factor necessary to elicit a potent cellular response.

An early study exploring this design concept focused on epidermal growth factor, which is effective in repair of damaged tissues, but is often difficult to deliver at a sufficiently high concentrations to mediate downstream signaling events since it does not contain a matrix-binding domain and rapidly undergoes receptor-mediated endocytosis. Immobilization of EGF on a biomaterial surface in a manner that preserved its activity but precluded receptor-mediated downregulation was first explored as a way to achieve high local concentration and sustained signaling in primary rat hepatocytes. In a recent example involving human and porcine mesenchymal stem cells (MSCs), amine-targeting chemistry was used to tether EGF to the surface of poly(methyl methacrylate)-graft-poly(ethylene oxide) (PMMA-g-PEO) comb polymers. The tethered EGF led to sustained EGFR signaling and subsequent cellular responses including cell spreading and protection from apoptosis whereas saturating levels of soluble EGF did not.

Motivated by naturally-occurring heparin-binding domains, Sakiyama-Elbert and colleagues incorporated heparin into biomaterial scaffolds to allow for immobilization of basic fibroblast growth factor (bFGF), a growth factor with many important roles during development and in adult physiology. bFGF was released either passively by diffusion, or actively via heparinases secreted by neighboring cells, thereby allowing for a controlled release and presentation of signal not possible with soluble growth factor delivery.

In a chick dorsal root ganglion model, immobilized bFGF led to greater neurite extension into the scaffold than for cells cultured on scaffolds with free bFGF, likely due in part to the finding that the local concentration in the scaffold was 500-fold greater for immobilized bFGF than for soluble FGF. The same delivery system has been used for differentiation of murine ESCs into mature neural cell types, including neurons and oligodendrocytes, indicating that biomaterials scaffolds functionalized with immobilized growth factors may be a potential strategy for generation of engineered tissue for treatment of spinal cord injury.

Finally, in a recent study, polymer substrates functionalized with the signaling domain of Sonic hedgehog supported enhanced osteogeneic differentiation of bone marrow-derived MSCs, as compared to cells cultured on the same surfaces with soluble Shh at the same concentration. This example further demonstrates that growth factor or morphogen immobilization serves as an effective means to achieve sustained activation of downstream signaling pathways.

Sequential Factors to Program Stem Cell Differentiation

Decades of elegant studies in numerous model organisms have elucidated signals and molecular pathways that sequentially and progressively regulate the generation of terminally differentiated cell types during the process of organismal development. One of the major accomplishments of embryonic stem cell biology thus far has been to apply this concept of a step-wise developmental program *in vitro* to recapitulate organ formation. In particular, generation of functional motor neurons and pancreatic cells are illustrative examples of how sequential soluble factors can be used to program stem cell differentiation.

Differentiation of motor neurons from primitive ectoderm requires progression through two intermediate stages, each guided by a distinct set of soluble factors. Ectodermal cells first acquire a rostral neural fate through signaling by BMP, FGF, and Wnt proteins. The next transition is retinoic acid-mediated caudalization, followed by progression to terminally differentiated motor neurons in the presence. In a landmark study, Jessell and colleagues were able to guide mouse ESCs down this pathway by applying these soluble factors in a step-wise manner that emulates natural neural development, resulting in the *in vitro* generation of motorneurons that can survive and engraft *in vivo*.

Analogous sequential exposure to soluble factors has enabled recapitulation of pancreatic development *in vitro*. The pancreas develops from the embryonic endoderm, first appearing as buds at two sites of the gut tube. Progression to the next stage of development, denoted by expression of pancreatic-specific transcription factors, depends on Shh-antagonists secreted by the embryonic mesoderm. Further growth and branching of the nascent pancreatic buds is mediated by FGF10 .

This step is followed by endocrine cell specification, which is mediated by soluble signals from the surrounding endothelium that inhibit Notch signaling and subsequently initiate expression a cascade of transcription factors that lead to differentiation of insulin-producing cells. Application of these soluble signals in sequence to undifferentiated embryonic stem cells *in vitro* resulted in the generation of functional insulin-producing cells, illustrating an important concept for directed differentiation of ESCs.

Factors Secreted or Presented from Differentiated Cells within the Niche

Stem cells reside in complex and heterogeneous compartments throughout the body, and interactions with one another and with surrounding cells can modulate their fate decisions. Therefore, the specific balance of cell types within the niche is a regulator of stem cell behavior. In some regions of the adult brain, signaling feedback from differentiated cells inhibits the proliferation of progenitor cells in order to maintain the balance of cell types in the niche. Wu et al. identified a TGF-β family member that inhibited neurogenesis in the olfactory epithelium, and then showed that it was produced by differentiated neurons. This system demonstrates the concept that differentiated cells can express a soluble protein to exert negative feedback control over stem cell proliferation, thereby maintaining a balance of signals and cell types within the local microenvironment.

Similar paracrine signaling has been shown to regulate human embryonic stem cell (hESC) cultures. Even when not grown on embryonic fibroblast or other feeder cells, hESCs grow in heterogeneous colonies composed of undifferentiated cells and hESC-derived fibroblast-like cells that do not express ES cell markers. Recent work suggests that the fibroblast-like cells form a part of the hESC microenvironment by secreting factors necessary for stem cell self-renewal. Specifically, fibroblasts derived from hESCs were shown to secrete insulin-like growth factor (IGF) II, which interacts with the IGF receptor I expressed on the surface of undifferentiated hESCs.

A closely related parameter is overall cell density. As one example, the Notch signaling pathway mediates the process of lateral inhibition, where a cell presenting a Notch ligand stimulates Notch activity in adjacent cells, which, in addition to modulating the recipient cell's fate, also downregulates the expression of Notch ligands on that cell. This signaling mechanism regulates key stages of neural development in both vertebrates and invertebrates. For example, specification of neuronal ganglion cells in the retina from a population of competent progenitors is mediated by Notch signaling such that inhibition of Notch leads to differentiation. A similar juxtracine signaling relationship is observed in *in vitro* neural stem cell cultures in which high cell density exerts an inhibitory effect on neuronal differentiation, which can be abrogated by the addition of antisense oligonucleotides directed against Notch.

One strategy to engineer paracrine and potentially juxtracrine regulation into a synthetic stem cell environment is to culture stem cells alongside more differentiated cell types. While these co-culture systems serve as the *in vitro* analog of native environments comprised of multiple cell types, they are often difficult to design and optimize, as well as to precisely control for basic investigation of paracrine signaling. Recently, tissue engineers have developed technologies that address some of the challenges of co-culture systems. For example, Hui and Bhatia used microfabricated, interdigitating silicon combs to control cell-cell interactions over time. In this system, cells are cultured on silicon combs whose spacing can be modulated over time. Micromechanical control of comb spacing allows for the modulation of cell-cell contacts and spacing to explore the roles of contact-mediated versus soluble signals among heterogeneous cell types.

For example, dynamic regulation of the spacing between hepatocytes and stromal cells revealed that stromal contact is only needed at an early time interval to sustain hepatocellular function. Similarly, Wright et al. used a microstencil to pattern various cell types layer by layer over time (Wright et al., 2007). This system allows for the optimization of each layer for each specific cell type (e.g. adhesive substrate coating), as well as the addition of cells in the co-culture in a defined sequence. Each of these systems represents emerging technologies that are a step towards recreating and exploring the complex dynamic cell-cell interactions that occur in stem cell niches.

Another approach to regulating cell-cell contacts in synthetic stem cell niches has been to regulate overall cell density. Negative feedback mechanisms mediated by cell-cell contacts can serve to maintain the balance of cell types in some stem cell niches, and controlling cell density can thus be considered a simpler proxy to recreating complex cell-cell contacts. One example comes from embryonic stem cell culture. In order to begin to parse the complex cell-cell interactions in hESC culture, Zandstra and colleagues have studied the effect of local cellular microenvironments on intracellular signaling pathways (Peerani et al., 2007).

Using high-resolution imaging, they showed that hESCs retain stem cell marker expression upon growth factor removal in regions of high local cell density within hESC colonies. Nuclear levels of signaling molecules such as phosphorylated Smad1, a component of the TGF-â pathway known to function in hESC self-renewal, were also found to be dependent on local cell density. Further analysis suggested that differentiated cells secrete Smad1 agonists, while undifferentiated cells secrete Smad1 antagonists. Moreover, using engineered surfaces with micropatterned cell adhesive features, they showed that controlling hESC colony size can be used to modulate the level of Smad signaling and maintenance of self-renewal. Their findings suggest that self-renewal requires a balance between TGF-β agonists and antagonists, and that this balance is determined by the subpopulations and densities of cells that comprise the local cell microenvironment.

This study indicated that ESC colony size can impact cell fate decisions, suggesting the importance of controlling this parameter. One technology to modulate ESC colony is the Bio Flip Chip (BFC), a polymeric chip fabricated with microwells that allow for cell patterning on culture substrates or other cell layers. Cells are pipetted on the surface of the BFC and allowed to settle into microwells. The chip is then inverted onto the desired culture substrate, allowing the cells to settle, attach, and grow in the desired pattern.

By altering the spacing between microwells, the chip can be used to investigate the effects of cell density and cell-cell contact on stem cell culture. Using this technology, Rosenthal et al. found that seeding mouse ESCs as single cells maximized the number of resulting undifferentiated colonies. This behavior is distinct from that of human ESCs, which have low survival rates as single cells (Stewart et al., 2006). Therefore, these studies demonstrate that modulating cell density can impact and potentially control cell fate, but in a cell-specific fashion.

Cell density is an important variable not only for the self-renewal but also for the differentiation of embryonic stem cells. The first steps of lineage commitment of embryonic stem cells can occur in embryoid bodies (EBs), which express markers of all three embryonic germ layers. EB formation is typically conducted in suspension culture via growth factor removal, which results in spheroids with a broad size distribution. The heterogeneity in EB size can impact further downstream differentiation through cell density and diffusion effects, among others.

To allow for more precise control over embryoid body formation, Karp et al. created microfabricated polymeric microwells for ESC differentiation that allow for control over EB size through variation of microwell dimensions (Karp et al., 2007). Murine EBs generated using this method had more uniform expression of á-fetoprotein, brachyury, and nestin – markers of the three embryonic germ layers – a result that may impact the efficiency of downstream differentiation processes.

DIRECT CELL-CELL CONTACT

Stem cells have important interactions with the heterogeneous cell types that comprise stem cell niches, among them direct cell-cell signaling relationships through molecules expressed on the cell surface. Presentation of the signaling molecules in a fluid lipid bilayer differs from ECM-immobilization, for instance, in that it allows for diffusion and receptor clustering. Clustering of cell surface receptors often serves to polarize cells and in some cases is necessary for activation of downstream signaling pathways.

Integrins, the major class of cell surface adhesion proteins, are well known to mediate signaling through receptor clustering. Phosphorylation of downstream signaling targets depends on β_1 integrin clustering (Kornberg et al., 1992), and presentation of synthetic integrin ligands in clusters reduced the ligand density required for cell adhesion (Maheshwari et al., 2000). Similarly, antibody-induced oligomerization of soluble, recombinant Notch ligand Delta is in many cases necessary for receptor binding, internalization, and activation of downstream signaling targets and has been used to activate Notch in neural crest stem cells. Presentation of membrane-bound signals is also a mechanism by which cell polarity is established in stem cell niches.

Study of the development of the germline in *Drosophila* has led to major insights into the role of the niche in directing cell fate decisions, and in particular the signals provided by neighboring cells. In the male germline stem cell (GSC) niche, one role of hub cells is to establish stem cell polarity (Yamashita et al., 2005). Stem cell polarity is necessary for asymmetric cell division, which is the primary mechanism of self-renewal. Adherens junctions between GSCs and hub cells enable orthogonal orientation of the stem cell mitotic spindle with respect to the hub, an elegant illustration of how niche cells guide stem cell fate decisions (Yamashita et al.,2003).

These examples of how membrane-bound or lipophilic factors can regulate cell function are particularly instructive for the design of stem cell microenvironments. As shown above, soluble ligands often do not result in the desired downstream effects. Thus, one successful approach in designing synthetic microenvironments for stem cell culture has been to incorporate ordinarily lipid-bound signaling molecules into the surface of biomaterials, thereby mimicking the native cell-cell interactions.

The morphogen Shh has important roles in development, and is thus a critical factor in stem cell engineering efforts. Lipid modifications at both the N- and C-termini of Shh enable interactions with the transmembrane protein Dispatched that lead to Shhmultimerization (Kawakami et al., 2002; Zeng et al., 2001). Motivated by the biochemistry of naturally occurring Shh, immobilization and/or oligomerization of Shh molecules on biomaterials in a manner that allowed for multivalency was explored as a technique to stimulate angiogenesis in a chick corioallantoic membrane assay (Wall et al., 2008). Multivalent, immobilized Shh had a more potent angiogenic effect than soluble Shh, illustrating how multimerization of lipophilic signaling molecules can be effectively emulated *ex vivo*.

The Notch signaling pathway is also broadly implicated in the regulation of numerous stem cell populations (Carlson et al., 2007); therefore, there is considerable interest in modulating the activity of this pathway in culture to regulate cell fate. Liu and colleagues have designed artificial ECM proteins that express the bioactive domain of Notch ligands Jagged and Delta for use in neural stem cell culture (Liu et al., 2003; Liu et al., 2003).

Similarly, microbeads functionalized with immobilized Notch ligands were used to direct differentiation of bone marrow-derived hematopoietic stem cells into T cells (Taqvi et al., 2006). Addition of the functionalized beads to an established lymphoid differentiation co-culture system shifted the balance of differentiated cells from exclusively B cells in the absence of Notch signaling to a mix of B and T cells, indicating that recreating the signaling environment normally provided by neighboring cells can recapitulate the niche *ex vivo*. This system can also be used to study effects of ligand density on downstream cell fates, providing additional design criteria for stem cell engineers.

Synthetic membranes have also been explored as a method to recreate cell-cell interactions *ex vivo*. In these systems, cells are cultured on a supported lipid bilayer presenting a signaling molecule. Using this technique to explore cell-cell interactions at the neuronal synapse, Groves, Isacoff, and colleagues report that clustering of signaling molecules in the membrane lead to strengthening of cell-membrane adhesion (Pautot et al., 2005). This work reconstituting the neuronal synapse with a synthetic lipid bilayer can be extended to stem cell applications.

ECM PRESENTS IMPORTANT BIOCHEMICAL AND MECHANICAL CUES TO STEM CELLS

The specific cell-cell interactions and growth factors that influence stem cell fates *in vivo* exist in combination with signals from the extracellular matrix. ECM has been primarily understood as the adhesive substrate that anchors cells within their microenvironments. For example, in the *Drosophila* GSC niche, abrogation of hub cell adhesion to the ECM resulted in mislocalization of hub cells and subsequent disruption of the mechanisms by which hub cells support GSC self-renewal (Tanentzapf et al., 2007).

However, the ECM itself can also provide instructive cues for cell fate decisions, primarily via the integrin family of cell surface adhesion receptors. In erythropoiesis, for example, adhesion of primary erythroid progenitors to the niche ECM protein fibronectin mediated by $\alpha_4\beta_1$ integrin is necessary for proper proliferation *in vitro* (Eshghi et al., 2007). In this system, signals from the ECM thus cooperate with signals from the soluble factor erythropoietin to activate signaling pathways necessary for terminal differentiation and proliferation.

ECM molecules are typically quite large and present diverse domains for cell adhesion. In addition, many ECM proteins such as fibronectin and laminin are characterized by alternative splicing events as well as post-translational modifications, making it sometimes difficult to work with purified proteins. Synthetic ECMs, comprised of polymer backbones functionalized with adhesion motifs, have thus been utilized to dissect the mechanisms by which ECMs support stem cell fate decisions. For example, the ECM molecule laminin, which is important in both *in vivo* and *in vitro* neural stem cell (NSC) niches, engages at least 7 distinct integrin receptor pairs (Powell and Kleinman, 1997).

In order to achieve greater control over NSC fate, synthetic ECMs incorporating several laminin motifs were designed and used to elucidate the differential effects of adhesion molecule motifs on stem cell fate decisions (Saha et al., 2007). This work demonstrated that, surprisingly, the canonical integrin adhesion sequence RGD was sufficient to support the proliferation and differentiation of NSCs, whereas other laminin-derived adhesion motifs were unable to support these stem cell functions.

The ECM in a given tissue typically exists as a mix of several different proteins – often collagens, laminins, and fibronectins – and high-throughput approaches have accordingly been developed to investigate the combinatorial effects of ECMs on stem cell fates. One system analyzed the combinatorial effects of ECM on differentiation of ESCs into hepatocytes, and it found that an approximately 140-fold difference between the least and most efficient conditions, suggesting that ECM components may strongly influence hepatic differentiation (Flaim et al., 2005). This work has been extended to include synthetic ECMs in addition to various soluble growth factors, yielding the interesting result that the effect of a given growth factor often depended on the substrate background in which it is presented (Nakajima et al., 2007).

In addition to biochemical motifs or signals, the ECM also provides a physical framework in which stem cells reside and develop. Importantly, the diverse tissues of the adult body exhibit a range of matrix stiffness spanning several orders of magnitude, and such differences in substrate stiffness have long been known to influence cell fate decisions in differentiated cell types. An emerging area of study in stem cell biology and engineering is investigation of the role of these mechanical cues in stem cell fate decisions.

Because mesenchymal stem cells can differentiate *in vitro* into cell types from tissues ranging from muscle, bone, and potentially brain, Engler and colleagues hypothesized that the mechanical cues provided by the ECM are particularly instructive in lineage specification. Their study provides evidence that matrix elasticity influences differentiation of human MSCs into osteogenic, myogenic, and neurogenic cells (Engler et al., 2006). Specifically, by culturing naïve MSCs on elastically-tunable polyacrylamide gels, they show that gels of bone-like stiffness led to osteogenic differentiation, gels of muscle-like stiffness resulted in myogenic differentiation and soft gels of brain-like stiffness led to neuronal differentiation.

This landmark work illustrates the principle that stem cells are regulated by the mechanics of the cellular microenvironment. Mesenchymal stem cells generate lineages that are present in load-bearing tissue, and it would therefore be interesting to determine whether this important principle extends to other stem cell types, including ones from tissues that are relatively mechanically insulated. It has recently been found that substrate stiffness collaborates with soluble medium conditions to regulate the proliferation and differentiation of adult neural stem cells (Saha et al., submitted). Cells exhibit an optimum in proliferation (inFGF-2) and an optimum in neuronal differentiation (in retinoic acid) at an intermediate stiffness that is characteristic of brain tissue. Furthermore, under conditions that induce general or nonspecific cell differentiation (serum), stiff substrates support the differentiation of GFAP-expressing astrocytes, whereas soft substrates preferentially support the differentiation of â-III-tubulin expressing neurons. Taken together, these results suggest that the design of *ex vivo* stem cell culture systems should consider mechanical cues in the microenvironment such as matrix stiffness as factors in guiding proper lineage specification.

BIOMIMETIC THREE-DIMENSIONAL MICROENVIRONMENT FOR CONTROLLING STEM CELL FATE

Since the potential of stem cells for therapeutic application was discovered by Evans and Kaufman, stem cells have been anticipated to treat cancer, type 1 diabetes, Parkinson's disease, Huntington's disease, coeliac disease, cardiac failure, muscle damage, neurological disorders and many other conditions. Stem cell-based therapies may represent the next generation of biologics to treat or cure many diseases for which adequate therapies do not yet exist. Currently, over 500 cell-therapy-based companies worldwide are translating the experimental research into clinical therapeutics. The example of a re-engineered trachea transplanted into a patient in Spain in 2008 has demonstrated the great opportunities for applying stem cells to the repair of degenerative tissues.

Nevertheless, before stem cell-based therapies are applied in clinics, stem cell behaviour upon transplantation should be precisely controlled and the mechanisms of stem cell interactions with its microenvironment need to be elucidated. Upon transplantation, stem cells and their derived lineages experience a multitude of biochemical, structural, mechanical and stimulatory cues that influence cell behaviour. For example, a fundamental understanding of the implications of the interplay between stem cell microenvironment components or stem cell niche factors (growth factors, cell–cell contact and cell–matrix interaction) and external forces will enable better control of therapeutic cells and effective regeneration of functional tissues.

The unique function of stem cells inside the human body can be achieved possibly by the specialized, three-dimensional microenvironment that surrounds them in native tissues. The microenvironment components have to be placed to their anatomical and functional locations; the interactions between those components are crucial for regulating stem cell functions. When stem cells are removed from their microenvironments, their functionality, phenotype and responses to environmental cues can often be altered. For example, muscle stem cells grown on the standard tissue culture plastic support have been found to lose their regenerative potential.

This poses great uncertainty in the clinical application of stem cells when the appropriate characterization, manipulation, proliferation and physio-chemical environment control are not adequately performed. On the other hand, the risk associated with tumorigenesis has been demonstrated by the case in which a 13 year old boy with a hereditary neurodegenerative disease developed a multifocal brain tumour after being treated three times with stem cell implantations in Moscow.

The potential risks highlight the gap that still exists between our current knowledge of stem cell performance *in vivo* in the niche and its intractability outside of that niche. To regulate stem cell differentiation into the right phenotype, an appropriate microenvironment should be created in a precisely controlled spatial and temporal manner. The creation of such a microenvironment requires applying engineering design principles and emerging technologies to mimic the complex three-dimensional biological structure. Engineered biomimetic microenvironments have been designed to regulate the balance between stem cell differentiation and self-renewal.

Comprehensive reviews have addressed the underlying biological principles driving the biomimetic microenvironments. In contrast, this review highlights the progress in creating microenvironments from an engineering perspective. This paper will discuss engineering methods for identifying and controlling the release and delivery of soluble factors, analysing and mimicking the extracellular matrix (ECM), probing cell–cell interactions and applying mechanical stimulation. The emerging techniques for creating biomimetic microenvironments are highlighted.

CONTROLLING STEM CELL FATE

Inside the human body, stem cells are surrounded by niche cells and embedded in an ECM, which defines the geometric configuration, signalling pathways and biomechanical characteristics of the microenvironment. Stem cell functions are determined by a variety of biochemical, structural, hydrodynamic, mechanical and electrical cues at spatial and temporal levels. The stem cell niche consists of a group of supporting cells, ECM and soluble environment factors at the specific sites. Cells respond to their immediate microenvironment wherein remodelling consequently occurs, via homotypic or heterotypic interactions with neighbouring cells, and with the tissue matrix.

The components contributing to the stem cell niche can be classified into:

- Soluble factors secreted by the stem cells or niche cells, present in the surrounding tissue or culture media;
- ECM or cell substrate;
- Direct cell-to-cell interactions to elicit cellular signals; and
- External mechanical and electrical forces such as fluid-induced shear stress, dynamics tensile and compressive loading.

The regulation of stem cell activities by its microenvironment has been validated in a number of different systems, such as germline stem cells from*Caenorhabditis elegans* and *Drosophila melanogaster*, and haematopoietic and neural stem cells from mammalian organisms. These elements function as a physical anchor to constrain stem cells, to adjust the concentration of extrinsic factors, and to regulate the intracellular signalling pathways. Thus, after cell division, one daughter cell is maintained in the niche while the other one migrates out of the niche to the target site and differentiates into a functional mature cell.

Although progress has been made to elucidate the mechanisms for regulating stem cells in their microenvironment in different systems *in vivo*, the highly dynamic and complex structures of stem cell niches will be revealed through elegantly engineering biomimetic microenvironments *in vitro*. More importantly, the engineering system, through accurately controlling physical and molecular interactions, enables directed regulation of stem cell behaviour, which extends our capabilities in engineering functional tissue substitutes *in vitro*. The engineering approach for mimicking the four key components of the stem cell microenvironment.

Soluble Factors

Soluble factors have been demonstrated to regulate stem cell growth and differentiation, including growth factors, morphogenetic factors, cytokines, enzymes and small cell-permeable molecules, such as basic fibroblast growth factor (bFGF), members of the transforming growth factor-β family (TGF-β), vascular endothelial growth factor (VEGF), bone morphogenetic factor (BMP), vitamin C, sodium pyruvate, retinoic acids (RAs) and other small molecules. For instance, bFGF plays a significant role in several pathways, including Akt and mitogen-activated protein kinase (MAPK) for a number of stem cell lineages, and all-*trans* RA is a strong differentiation agent to enhance expression of the neural crest and reduces mesodermal differentiation. The nutrient components including oxygen are also considered as soluble factors in the biomimetic system.

These factors added to the cell culture system or secreted by stem cells or niche cells are potent in their effects on stem cell fate. For example, cytokine leukaemia inhibitory factor (LIF) is able to maintain the self-renewal capacity of embryonic stem cells (ESCs) in the presence of serum without embryonic fibroblast feeder cells. It has also been noticed that the combined effect of these factors can vary substantially. For instance, co-presence of BMP4 and LIF in serum-free media promotes ESC self-renewal, while BMP4 alone induces ventral mesodermal differentiation.

The above addition method in the cell culture media provides clues to the biological functions of soluble factors in the microenvironment; however, it cannot control the delivery of these factors in a spatial and temporal way. Several approaches have been employed to design the spatial concentration gradient of soluble factor(s) of interest. Among these, microfluidic technology is an emerging method to generate molecular gradients in soluble form through laminar flow within the microfluidic channel. Microfluidics enables massive and parallel manipulation of a tiny volume of solutions containing multiple soluble factors or combinations in a multi-dimensional space to determine stem cell function.

This approach is exemplified by a microfluidic alginate hydrogel controlling release of soluble factors in three-dimensional microenvironments. A lithographic technique was used to build functional networks within a calcium alginate hydrogel seeded with cells. Temporal and spatial control of the distribution of non-reactive solutes and reactive solutes (metabolites) within the bulk of the scaffold has been demonstrated. This approach can control the chemical environment on a micrometre scale within a macroscopic scaffold and is useful in engineering complex tissues.

In the biological system, these soluble factors are often bound to ECM to slow their diffusion and fine-tune their local concentrations and gradients. The biomimicking approach is to conjugate or bind soluble factors to the surface of biomaterials such as ECM, three-dimensional scaffolds, hydrogels, macro- or microporous foams and woven and non-woven fabrics in a spatial and temporal manner. Soluble factors released from biomaterials are normally controlled by diffusion, cell-mediated proteolysis and external physical stimulation.

For example, growth factors can be localized in multicomponent, spatially patterned and photo-cross-linked hydrogels. Human ESCs were encapsulated in a dextran-based hydrogel with two immobilized regulatory factors, a tethered (arg-gly-asp) RGD peptide and microencapsulated VEGF165. The fraction of cells expressing VEGF receptor KDR/Flk-1, a vascular marker, increased up to 20-fold, which demonstrated that this approach can be used to induce vascular differentiation of human ESCs.

Extracellular matrix

The ECM forms a complex architecture containing proteins, polysaccharides and proteoglycans, and these molecules constantly undergo dynamic change owing to assembly, remodelling and degradation events. The ECM not only provides architectural guides for tissue development, but also defines and maintains cellular phenotype and drives cell fate decisions. At a molecular level, the ECM has been demonstrated to influence stem cell fate by mechanical traction forces and cell adhesion. When cells anchor to the ECM, they exert inward-directed traction (tensional) forces generated by the contractile cytoskeleton because of adhesion to substrate surfaces.

The deformation induced by the traction force activates transmembrane integrin signal pathways that affect the cell fate. The magnitude of these traction forces depends on the mechanical properties of the ECM, such as stiffness. At cell adhesion sites, stem cells adhered to the specific component of the ECM via integrins, cadherins and discoidin, etc., activate unique signalling pathways. The adhesion also determines orientation of the stem cell division plane and cell morphology through the constraints imposed by the surrounding ECM, and consequently influences stem cell self-renewal and differentiation.

Identification of Extracellular Matrix Components

The identification of ECM molecules and their roles in stem cell regulation are critical steps towards creating biomimetic microenvironments. Combinations of specific molecular interactions between numerous isoforms, ratios and geometrical arrangements of collagen, elastin, proteoglycans and adhesion proteins (such as fibronectin and laminins) for identifying and mimicking the natural ECM composition are extremely challenging. Furthermore, the composition of the ECM is normally tissue specific. For instance, laminin is the major component in basement membranes, while stromal ECMs in connective tissues consist mainly of collagen.

The ECM composition is also dynamic, varying as the tissue is being developed. Fibronectin-rich ECM can promote proliferation of immature capillaries, while as the capillary matures, the composition of the ECM changed to become laminin-rich. To screen an ECM component or combinations, novel methods, such as robotic spotting and microfabricated wells, have been developed to make protein arrays. In a single experiment, typically more than 1000 combinations of ECM molecules can be screened simultaneously.

For example, Flaim *et al.* prepared an ECM microarray that combined five different ECM molecules in varying ratios and demonstrated that combinations of ECM components were more effective than single ECM components for differentiating mouse embryonic stem cells (mESCs) into liver cells. Anderson *et al.* developed arrays consisting of 192 unique combinations of ECM. Signalling factors were printed onto slides containing a thin coating of polydimethylsiloxane, and ECM molecules were identified to maintain the progenitor state, and to guide progenitor differentiation towards myoepithelial and luminal lineages.

Functional components of ECM *in vivo*. However, natural ECMs are variable in their composition and mechanical properties. Naturally derived or synthetic gels or scaffolds have been tailored to mimic specific ECMs for the creation of well-controllable and reproducible microenvironments. Naturally derived three-dimensional structures can be from protein-based (collagen, fibrin, silk, gelatin and Matrigel), or from polysaccharide-based (hyaluronan, alginate, agarose and chitosan), or protein- and polysaccharide-combined decellularized matrices.

Synthetic materials include poly(ethylene glycol) (PEG), poly(ethylene oxide) (PEO), polyacrylamide, poly(vinyl alcohol) (PVA), poly(L-lactic acid) (PLA), poly(glycolic acid) (PGA), poly(L-lactic-co-glycolic acid) (polyPLGA), poly(hydroxyl ethyl mathacrylate) (PHEMA), poly(anhydride) and functional peptide self-assembly three-dimensional structures. ECM-derived peptides or protein fragments need to be covalently linked to polymer backbones via their hydroxyl-, carboxyl- or amino-termini to improve their biological functions.

Ligand Presentation of Extracellular Matrix

Apart from the composition of natural ECMs, the surface and mechanical properties, such as ligand density, surface nanotopography and substrate elasticity also influence stem cell fate. The RGD peptide in various ECM molecules such as fibronectin and vitronectin is the prevailing adhesive ligand because the binding of most cells to ECM is dependent on RGD density. Other adhesive peptides, such as the YIGSR and IKVAV peptides of laminin and the VPGIG sequence of elastin, have also been studied. Cell behaviour can be influenced by the coexistence of multiple peptides. For example, a spacing of 4 nm between RGD and the synergy site pro-his-ser-arg-asn (PHSRN) resulted in an increase in indicators of osteoblastic cell function, such as metabolic activity and alkaline phosphatase production, while showing a decrease in ECM production. Although some adhesive ligands have been integrated with PEG-based hydrogels, the incorporation of ligands into artificial ECMs and control of the distribution of those ligands on the surface in a spatio-temporal manner are still very challenging.

Extracellular Matrix Nanotopography

Nanotopography of the ECM has a great impact on cell adhesion. Not only the scale of topography (5 nm to micrometre scale) but also the type of ordered topography, such as ridges, steps, grooves, pillars and pits can modulate cell behaviour. Ruiz & Chen micropatterned fibronectin in various geometries (circle, square, rectangle and ellipses), and measured the tractive force experienced by human mesenchymal stem cells (hMSCs) on these fibronectin patterns. They concluded that cells on the concave surface experience greater tractive forces than those on convex surfaces, and the forces direct cells into osteogenic instead of adipogenic lineages.

The tractive forces also guide cytoskeletal formation, hence, to determine the cell shape. Evidence has suggested that physical control of cell shape alone can act as a potent regulator of stem cell fate. For example, single mesenchymal stem cells (MSCs) cultured on small micropatterned islands adhered poorly, displayed a rounded morphology and acquired an adipogenic fate, whereas those cultured on larger islands adhered strongly, spread out, exhibited increased focal adhesion and cytoskeletal reorganization, and acquired osteogenic fate. Mechanisms for the sensing of matrices as well as for stem cell trafficking are still to be clarified. The ability of cells to recognize the nano-scale topographic features requires developing synthetic ECMs with nano-structure. Self-assembly of amphilic peptides and electrospinning have been used to form nanofibres in biomimetic cell substrates.

Extracellular Matrix Mechanical Stiffness

The mechanical stiffness or elasticity of the ECM has a major influence on cell behaviours, such as migration, apoptosis and proliferation. Cells have been attached to various gel matrices with controllable elasticity by varying degrees of cross-linking and non-limiting ligand density, and most of the cells have been found to anchor more strongly to stiff substrates, building focal adhesions and forming actin–myosin stress fibres. MSCs cultured on the rigid substrate preferentially differentiated into osteoblasts instead of adipocytes, while muscle stem cells on soft hydrogel substrates at a similar elasticity of muscle maintained self-renewal *in vitro* .

CELL–CELL INTERACTIONS

Cell–cell interactions have been widely studied *in vivo* in different systems. They not only act as a physical anchor to constrain stem cells in a defined space, but also to provide instructive secreted signalling cues or to send signals through transmembrane proteins or bound matrix proteins. Interactions between cell populations can be divided into homotypic or heterotypic interactions. Homotypic interactions are between stem cells themselves, while heterotypic interactions are between stem cells and their neighbouring cells.

Stem cells communicate with niche cells through the tight junctions, adherens junctions, notch signalling pathways and gap junctions. Signalling molecules generated by the niche cells pass the channels between the stem cell and the niche cell to regulate the stem cell behaviour. Alternatively, they interact with each other via paracrine signalling, and diffuse soluble signal molecules from the neighbouring cells to the stem cells, such as Wnt, BMP, JAK/STAT pathways. In addition to the extensive *in vivo* study of cell–cell interactions, biomimetic approaches for these interactions have been explored to induce stem cell differentiation and promote expansion and self-renewal.

A common approach to investigating homotypic interactions is to plate cells with different cell seeding density. However, as the cell density is not able to be controlled locally, it is difficult to differentiate the effect of paracrine signalling from actual physical contact between the cells in the culture dishes. The effects of heterotypic interactions are usually investigated by placing two cell types in close contact or co-culturing them.

Nevertheless, particular signalling molecules generated by niche cells in co-culture methods may be neglected. A smart design to distribute cells in a micropattern allows for control of the cell–cell contact between cells adherent on RGD microdomains. In this way, the interference of soluble factors or cell seeding concentration can be overcome. Traditionally, the cell–cell interactions occur in two-dimensional cell culture systems, which may not be able to generate the same channels or signalling molecules as those in three-dimensional *in vivo*.

The study of cell–cell interactions in three-dimensional microenvironments *in vitro* is still a great challenge as it is difficult to precisely control cell–cell contact in a spatial dimension. To overcome this challenge, physical micromanipulation techniques have been used to accurately manipulate, move and organize cells in three-dimensional and to control cell–cell contact. For example, bioprinting can be used to generate spatially oriented co-culture by printing layer-by-layer bioactive molecules with a well-defined gap; laser-deposited human umbilical vein endothelial cells (HUVECs) self-assembled into vascular structures, while a combination of primary rat hepatocytes heterogeneously patterned with HUVECs resulted in tubular structures; mammalian cells were electropatterned within three-dimensional hydrogels with dielectrophoretic forces; and large numbers of multicellular clusters of precise size and shape were formed in three-dimensional on one focal plane. The cell–cell interactions were affected by clusters of various sizes: smaller clusters influenced the biosynthesis of bovine articular chondrocytes, while larger clusters produced smaller amounts of sulphated glycosaminoglycan per cell.

MECHANICAL STIMULATION

Mechanical forces play an important role in cell attachment, spreading, proliferation, migration and differentiation. External applied forces not only directly transmit to cells, but also change the relative distance between cells, ECM components and soluble factors.

Several molecular bases, not exclusive, for mediating cellular activity include:

- Activating ion channels by stretching;
- Inducing conformational change of proteins or unfolding of ECM proteins, leading to changes in protein activity;
- Inducing conformational change of cytoskeletal elements such as filaments, cross-linkers or motor proteins;
- Direct action on gene expression owing to forces transmitted to the nucleus; and
- Changing the intercellular space by stretching or compression, leading to changes in the local concentration and gradient of secreted signal molecules, and ECM ligand presentation and nanotopology.

Local variations of ECM elasticity at the micro-scale result in different abilities to resist cell traction forces, and thereby regulate cell and tissue development *in vivo*. It is difficult to precisely determine the biomechanical and biochemical mechanisms *in vivo* owing to the above inter-connected molecular changes and also dynamic cellular mechanical properties. Biomimetic systems have, therefore, been developed to characterize stem cell responses to more highly controllable mechanical stimulation and to determine the biophysical mechanisms and biochemical signal transduction pathways. Biomimetic mechanical stimulation includes deformation loading (compression, stretching and tension) and fluid-induced forces (pressure or shear stress).

Mechanical Strains

Mechanical strain from applied loading, either cyclic or uniform biaxial, can direct specific stem cell lineages. For example, cyclic stretch has been demonstrated to commit MSCs to a myogenic phenotype and mESCs to a vascular smooth muscle cell phenotype. Cyclic compression is able to alter MSC phenotype as well. MSCs subjected to dynamic compression or hydrostatic pressure showed increased chondrocyte lineage differentiation and enhanced ECM deposition. Mechanical strain can also increase proliferation and inhibit differentiation in mouse and human ESCs as well as foster cell alignment with respect to the direction of strain.

Fluid-induced Shear Stress

Fluid-induced shear stress (FISS) has a significant impact on the fate of stem cells as well. In two-dimensional systems, FISS induces osteogenic differentiation of stem cells by activating multiple intracellular signalling pathways, such as the signalling pathways of nitric oxide (NO)/cyclic guanosine monophosphate-dependent protein kinase (PKG), prostaglandin E2 (PGE2)/cyclic adenosine monophosphate-dependent protein kinase (PKA), Ca^{2+}/calmodulin-dependent protein kinase (PKC) and MAPK.

Interestingly, while expression levels of osteogenesis-related genes, alkaline phosphatase (ALP), osteopontin (OPN), collagen type 1 (Coll1), PGE2 and NO have been promoted in multiple studies, some other studies showed statistically insignificant changes in the levels of ALP, Coll1 and PGE2. The contradictory conclusions may be due to the variation in cell types used, culture conditions, and experimental platforms.

Stem cells may also be differentiated towards endothelial cell fate in response to shear stress via the VEGF signalling pathway. Most of the studies showed fluid shear stress can direct ESCs or endothelial precursors towards positive endothelial differentiation; however, shear force alone seems insufficient to differentiate adult MSCs or human adipose-derived stem cells into endothelial cells .

In three-dimensional culture systems, both osteogenic and angiogenic differentiation have been induced by FISS for stem cells in a range of 1×10^{-4} to 1.2 Pa, and the majority of work was focused on in a range 0.01–0.05 Pa, which is at least an order of magnitude below the average shear stress for two-dimensional culture and up to two orders of magnitude lower in some cases. hese values are also orders of magnitude below those expected to cause differentiation *in vivo*. This may be due to morphological deformation when attaching to the three-dimensional surface. However, these findings are not only dependent on the magnitude of FISS, but also on the soluble environment, cell–ECM interactions, three-dimensional surface topology and material stiffness, as these will alter cell focal adhesions and stem cell colony shape. For example, a low flow rate resulted in cell clumping on aggregation supporting embryo body culture owing to a lower rate of mass transfer to cells.

SYNERGISTIC EFFECTS

The above individual niche factors rarely function alone in the stem cell microenvironment, which further complicates the elucidation of the mechanisms of controlling stem cell fate. For instance, soluble factors first diffuse through the ECM and then reach the cell membrane surface. The ECM structure may block or delay the delivery of these soluble factors. It is of great interest and importance to investigate the synergistic effects by optimally combining those niche factors. However, this will require multi-factorial experimental designs and sophisticated statistical analyses.

The synergistic effects can be caused by a combination of the same or different categories of niche factors. For example, MSCs were exposed to a combination of three stresses: a pulsatile pressure, radial distension and FISS. They exhibited a similar mechanosensitive response to that of endothelial cells. However, gene expression results show the cells expressed greater levels of smooth muscle cell-associated markers. This result highlights the synergistic effects of physiological flow and stretch on cell behaviour. However, this also increases the complexity of characterizing the individual mechanical force.

A mathematical modelling approach may be integrated with the experimental methods, so the synergistic effects of biological phenomena can be quantitatively correlated with the individual contributing factors. Another example from Salvi *et al.* combined the mechanical stimulation and the ECM topology. They investigated the effect of FISS when MSCs were seeded and cultured on randomly distributed nano-island surfaces with varying island heights. Their observation suggests that specific scale nanotopographies provide an optimal milieu for promoting stem cell mechanotransduction activity. The mechanical signals and substrate nanotopography may synergistically regulate cell function.

The soluble factors were also combined with mechanical stresses to determine stem cell fate. A combination of an electrical stimulation and soluble factors induced synergistic osteogeneric differentiation of MSCs, while use of electric stimulation alone fails to differentiate MSCs. Endothelial progenitor cells were differentiated into endothelial cells by using a combination of VEGF and shear stress, while chrondrogenic differentiation of MSCs was achieved by TGF-β3 and a hydraulic pressure. Once the mechanisms of the synergistic effects are well characterized, strategies to optimize the combination of the niche factors can be formulated to differentiate stem cells into the lineages of interest.

PERSPECTIVES AND EMERGING TECHNOLOGIES

From the above discussion, it demonstrates that cellular activities are mediated by a variety of molecular, structural, hydrodynamic, mechanical and electrical cues and combinations in a spatial and temporal manner. Before stem cells can be used for clinical applications, the interplay between these cues must be understood. It is strongly recommended that an engineered biomimetic three-dimensional system should be developed to closely mimic the human system to correctly promote differentiation or proliferation of stem cells. Such an elaborate system would range from nano-scale to millimetre-scale, involve multiple cues and combinations, and require multi-disciplinary collaboration. Four future trends in designing the biomimetic system, which can be envisaged are reported here.

SYNTHETIC BIOMATERIALS AS INSTRUCTIVE EXTRACELLULAR MICROENVIRONMENTS

A biomaterials approach has been employed to identify the composition of the stem cell ECM and to synthesize biomimetic three-dimensional scaffolds or hydrogels. However, well-defined synthetic three-dimensional systems are far more challenging and require mimicking the mechanical and biological properties of the ECM, such as ligand presentation, nanotopography, substrate elasticity, growth-factor binding, degradation and remodelling. Furthermore, currently three-dimensional cell culture structures are normally prepared without cells. When seeding cells onto the three-dimensional system, cells are not uniformly distributed. Ideally cells are homogeneously suspended in the polymer solution, which will then form the three-dimensional materials.

A cell-compatible and highly specific cross-linking reaction will be required to maintain cell viability. DeForest's group has prepared a hydrogel from its monomers in the presence of cells using a 'click' reaction, so that the cells were encapsulated in the gel. When irradiated with light, the chemical group reacted in a second click reaction with molecules acting as signals that monitor or dictate the behaviour of the encapsulated cells. The *in situ* synthesis and remodelling of functional biomaterials to incorporate clusters of ligands and growth-factor binding sites will enable cells to evenly distribute within the three-dimensional network.

MICROSCALE TECHNOLOGIES FOR FINE TUNING AND SCREENING NICHE FACTORS

Microscale technologies have been extensively used to examine the three-dimensional microenvironment *in vitro*. They have two important features for stem cell studies. First, they allow fine tuning of stem cell niche factors at the scale of from 1 μm to 1 cm. Examples of employing microscale technologies are generating soluble factor concentration gradients in the laminar-flow microfluidic channel; creating micropatterning for confinement of individual cells to control the cell shape; and designing microbioreactors or microwells for investigating cell–cell contact and mechanical stimulation. Secondly, microscale technologies offer high-throughput platforms because they consume a minute amount of volume and operate automatically by integrating with a robotic liquid dispensing system. Owing to the complexity of the factors affecting stem cell differentiation, it is essential to analyse the stem cell microenvironment in a high-throughput manner. Microscale technologies for stem cell research and tissue engineering have been excellently reviewed by Khademhosseini *et al*. and Toh *et al*.. One can access the two reviews for more information.

MICROMANIPULATION TECHNIQUES FOR PROBING CELL–CELL INTERACTIONS IN THREE-DIMENSIONAL ARCHITECTURE

Three-dimensional manipulation techniques can be valuable for precisely controlling the three-dimensional architecture and cell attachment, and probing cell–ECM interaction and cell–cell interaction. These techniques include the use of contact-dependent atomic force microscopes, non-contact optical tweezers, dieletrophoretic traps and magnetic tweezers. Optical tweezers, for instance, have emerged as an essential tool for manipulating single biological cells and performing sophisticated biophysical/biomechanical characterizations without any contact. The ability to manipulate cells in a three-dimensional architecture will help to probe cell–cell and cell–ECM interactions in a biomimetic system.

Another way to probe cell–cell interactions is by bioprinting. Bioprinting is a rapid prototyping strategy to produce three-dimensional structures through computer-aided, localized deposition of multiple types of cells and biomaterials, which can create three-dimensional cellular microenvironments to mimic the natural physiological, geometrical, mechanical and regulatory cues. Currently, there are two common types of bioprinting: ink-jet printing (IJP) or laser-based direct writing (LBDW). By applying the LBDW technique, hMSCs have been directly written on a Matrigel-coated substrate using a 1064 nm wavelength laser. Two-dimensional patterns consisting of one or more cell types were generated and the cell viability was around 90 per cent. Similarly, mESCs were deposited into defined arrays of spots, and stem cell pluripotency was maintained.

This technique can also be integrated with CAD/CAM at a single cell resolution using a minute amount of volume. For example, HUVECs were direct-written in a three-line pattern on a collagen-coated surface. The media was aspirated and a collagen gel (0.5 mm high) was carefully layered on top of the first pattern. An additional pattern of three lines of cells was written and oriented perpendicularly to the first pattern on top of the gel to create a three-dimensional pattern. This technique also demonstrates the ability to place one cell on top of another in a true three-dimensional pattern, offering the potential to investigate the cell–cell interactions in three-dimensional. Bioprinting can also precisely construct the ECM layer-by-layer to form a well-defined biomimetic anchor for stem cells.

MATHEMATICAL MODELLING FOR QUANTITATIVE ANALYSIS OF CELL BEHAVIOUR AND MICROENVIRONMENT PARAMETERS

After an engineered three-dimensional microenvironment is built, it is essential to assess the performance of the microenvironment by cellular responses to individual components including cell growth, migration, differentiation or apoptosis, as well as to characterize the parameters associated with the engineered microenvironment. Cellular activities depend not only on the presence or absence of cues, but also on their quantity, their spatial arrangements and the temporal order in which they are presented. Parameters associated with the microenvironmental cues include the substrate elasticity, soluble factor concentration and distribution inside the microenvironment, mechanical stress on the cells and the ECM geometrical information (nanofibre size, porosity, surface chemistry, etc.). These parameters are unfortunately difficult to achieve. Computational models have been developed to characterize the engineering microenvironment as well as dynamic change in cellular activities.

The flow-related parameters, such as oxygen, soluble factor concentration and shear stress, can be characterized by computational fluid dynamics, which reveal the local flow information in the microenvironment in a complex geometry. The dynamic nature of the cellular activities can be described by cellular automata models. Both continuous and discrete models have been employed to capture tissue level dynamics with varying levels of success. Many of the latter type are emerging to explicitly model a large number of individual cells, which behave within a certain biological framework with respect to cell adhesion, division and migration. Future models will combine the above two models, accounting for force-dependent molecular switches, signal transduction pathways, cell deformation and tension and reciprocal interactions with the microenvironment.

Quantitative prediction through both multi-level (from molecules to tissue) and multi-scale (from nano- to millimetre) computational modelling can help to correlate the cellular performance with parameters of engineered microenvironments. It will play a critical role in elucidating the mechanisms for governing stem cell behaviour in the stem cell microenvironment.

BIOCHEMICAL CONTROL OF STEM CELL RESPONSE

Self-renewal and lineage commitment of SCs have shown to be influenced by several cues, which can be subdivided into two classes: biochemical cues and biophysical cues. Biochemical cues are provided by reciprocal interactions between the cell, soluble bioactive agents, and the ECM.

Soluble factors include growth factors, morphogenetic factors, cytokines, enzymes, and small cell-permeable molecules, such as transforming growth factors (TGF), bone morphogenetic protein (BMP), vitamin C, sodium pyruvate, retinoic acids (RAs), and other small molecules. These factors, when added to the cell culture, or secreted by stem cells or niche cells, diffuse and bind to cell membrane receptors activating cellular signal pathways able to alter SCs gene expression.

Regarding the biochemical cues provided by the reciprocal interaction between the cell and the ECM, a strategy that has been developed to modulate cell attachment, proliferation, and differentiation consists in the use natural biopolymers to build scaffold materials. Also synthetic biomaterials have been exploited. A property that has been shown to regulate SCs activity is chemical functionalisation of the substrate surface, for example, by anchoring monomers representing the ECM binding sites, including RGD and IKVAV, or other functional groups, for example, $-CH_3$, $-NH_2$, –SH, –OH, and –COOH.

Even though the effect of biochemical cues, including soluble factors and ECM ligands, has been widely investigated in vitro, the regulation of self-renewal and lineage commitment of SCs by these key factors is still poorly understood and difficult to mimic; actually, in living systems, ECM components play a crucial role in the controlled delivery of molecular signalling molecules. Furthermore, the spatial and temporal organization of ECM adhesive ligands in vivo is finely tuned and difficult to achieve through engineering methods.

Besides applying strategies to control SC fate using biochemical factors, there is increasing evidence that mechanical factors are potent enough to control their fate in vitro. For instance, it has been demonstrated that the matrix elasticity can influence the lineage commitment of MSCs into neurons, osteoblasts, and myoblasts. Other than the substrate stiffness, these mechanical factors include the surface topography of the biomaterial scaffold, its three-dimensional (3D) geometry, and the external forces applied to cells.

MECHANICAL REGULATION OF STEM CELL FATE

Cells have the ability to actively sense their microenvironment and to react to the properties of their surroundings. Anchorage-dependent cells are able to anchor onto the underlying substrate through focal adhesions formed by clusters of proteins, including integrins, that are trans-membrane cell adhesion proteins. The cell cytoskeleton, mechanically linked to the focal adhesions, is a network of filamentous proteins that extends throughout the cell cytoplasm in eukaryotic cells, and consists of actin, microtubules, and intermediate filaments.

The mechanical connections between the matrix and the cytoskeleton allow cells to exert traction forces that are transmitted to the cell nucleus through intracellular pathways; the resulting force triggers signalling transduction into biochemical signals that affect SC response, for example, the synthesis of specific transcription factors in the nucleus. Various mechanotransduction pathways have been proposed, including the Ras/MAPK, the PI3K/Akt, RhoA/ROCK, Wnt/β-catenin, and the TGF-β pathways, which are generally integrin-based, and mechano-sensitive ion channels. An extensive review of known mechanotransduction pathways can be found.

THE CONCEPT OF "FORCE ISOTROPY"

At the single cell level, cell adhesion to a material with specific engineering properties, such as surface nanotopography, stiffness, and microgeometry, can induce cytoskeletal tensional states due to the complex system of traction forces exerted by cells on the surrounding microenvironment; as mentioned above, these forces trigger signalling transduction cascades that provide intracellular events that regulate cell behaviour. The cytoskeletal tensional state at the generic time point depends on the balance between the intracellular actomyosin contractility and the reaction forces exerted by the underlying substrate, determined by its engineering properties.

If cell adhesion-mediated traction forces, exerted by the cell on the matrix or adhesion substrate, have different magnitudes at varying spatial orientations, as shown by arrows, then the cell nucleus tends to elongate; this condition is what we define as "anisotropic cytoskeletal tension". On the contrary, isotropic cytoskeletal tension is characterized by adhesion-mediated traction forces of similar magnitude at varying orientation; in this condition, the cell nucleus tends to maintain a roundish morphology.

In addition to the cytoskeletal tension related to cell adhesion, also extracellular forces are able to influence SC commitment ; extracellular forces, including pressurization of interstitial fluids, flow-induced shear, and matrix strain, are dynamically superimposed to the cell cytoskeletal tension in native tissue. The concept of force isotropy may well be extended to extracellular forces. When adhering cells are subjected to anisotropic loads, including cyclic tension or shear stress, or to isotropic loads, such as cyclic hydrostatic pressure, they will respond to these forces, which are superimposed to cytoskeletal tension, by dynamically varying the spatial distribution of focal adhesions, thus the nucleus shape. Increasing levels of anisotropy, both in terms of cytoskeletal tension and in terms of extracellular forces, can be obtained in vitro by applying engineering strategies.

ENGINEERING STRATEGIES TO MIMIC FORCE ISOTROPY STATES

The ability of various types of cells to respond to mechanical differences in the extracellular environment has been thoroughly reviewed by Discher et al.. The different engineering strategies developed to study how the cytoskeletal tension and extracellular forces affect SCs response are summarized below.

Substrate Stiffness

A biophysical factor that has been proved to strongly affect adult SC behaviour is the substrates stiffness. By varying the mechanical properties of the matrix, for example from soft to relatively rigid by controlling the level of polymer crosslinking, it's possible to direct SC fate. Regarding MSCs, when they are cultured on soft substrates that mimic the elasticity of brain tissue (characterised by a stiffness of 0.1–1 kPa, in terms of Young modulus), neuronal precursors were generated; matrices with intermediate stiffness mimicking muscle (8–17 kPa) induced myogenic commitment while comparatively rigid matrices mimicking collagenous bone (25–40 kPa) proved to direct osteogenic lineage.

Substrate stiffness was found to also strongly influence ESCs behaviour. For example, it has been found that cell spreading increases as a function of substrate stiffness. In a further study, in which a wide range of substrate stiffness was investigated (41–2700 kPa), it was assessed that mouse ESCs cultured for 5 days on softer substrates were able to self-renew even in the absence of leukemia inhibitory factor (LIF).

Nanotopography

Another key factor that has been demonstrated to direct self-renewal and commitment of SCs is the surface topography at the micro- or nano-scales. Various patterns combined with different adult SC types have been investigated: nanostructured surfaces affect self-renewal; nanogratings of 350 nm width compared to flat substrates can induce an upregulation on neuronal markers of MSCs even in the absence of retinoic acid as a biochemical cue; neuronal commitment in human MSCs was also observed in nanofibrous scaffolds, with fibre diameter in the range of 230 ± 31nm, using neural induction factors.

Myogenic commitment of MSCs in the absence of differentiating medium has been observed in aligned/nonaligned nanofibres having different average diameters and in low roughness silk-tropoelastin scaffolds. Grooves and grids, inducing cells alignment and elongation, show a greater effect on osteogenic differentiation compared to flat surfaces in MSCs. Also random topographies have been exploited, including controlled disorder nanopits; the disordered ones were able to induce MSCs to differentiate towards osteogenic lineage. The height of topographical features may influence MSCs differentiation and adhesion. In this study, the authors demonstrated that SC behaviour is a function of pillar height and not of the diameter or gaps between nanopits.

The pattern shape has been proved to regulate MSCs activities: in round patterns aided adipogenic lineage differentiation while gratings induced elongation and osteogenic transition; also, in geometric-specific proliferation and differentiation patterns were observed. Regarding neuronal progenitor SCs, nanofibres of various size have been investigated and it was observed that cells grown on these fibres were able to differentiate into oligodendrocytes. The proliferation rate increased on smaller fibre diameter. Neural differentiation was observed on grooves chemically modified with laminin.

Nano/microtopography has shown to affect also ESCs behaviour. In mouse ES cell differentiation can be observed when these cells are cultured in the form of embryoid bodies. ESC morphology, including elongation, alignment, and proliferation can be induced by nanoscale grating features. Micro- and nanopatterns of various geometries and sizes to study ESCs response have been exploited; for example, in circular micro domains were found to direct cardiac differentiation of uniform-sized mouse ESC aggregates. Also nanofibres have been investigated and it has been shown that they induce an upregulation of early development osteogenic markers.

Microgeometry

The three-dimensional (3D) microgeometry of the culture environment strongly contributes to direct cell phenotype. Traditionally, mammalian and SCs have been cultured on 2D substrates. Differences in cell behaviour, including adhesion, migration, proliferation, and gene expression, have been observed between 2D and 3D cultures. For instance, 2D culture confines cells to a planar environment and restricts the more complex morphologies observed in vivo; cells are able to interact through a limited membrane segment with the underlying substrate and neighbouring cells. As a consequence, the mechanotransduction process and the interaction with nutrients, soluble factors, and mechanical cues are altered.

These considerations lead an increasing interest in the development of truly 3D scaffolds to mimic the native environment in which cells reside. Different 3D cell culture supports have been developed, including nanofibres, hydrogels, microwells, and other more complex structures. Nanofibrous scaffolds were able to mimic the architecture formed by fibrillar ECM proteins. Another approach to surround cells with a truly 3D environment consists in suspending them in a hydrogel; for example, in osteogenic commitment of rat MSC has been observed in these gels, although in the presence of osteoinductive medium. Very recently, photocurable materials have been used to fabricate 3D scaffolds by two-photon polymerization, a 3D microfabrication technique that allows to fabricate ordered microstructures having a submicron spatial resolution. Cell viability, homing, and proliferation were promoted by culturing MSCs on these structures.

External Forces

Mechanical forces play an important role in SC adhesion, spreading, proliferation, migration, and differentiation. External forces applied to the cell are able to influence the intracellular cytoskeletal arrangement thus regulating nuclear shape and cell activities. It is difficult to precisely determine the biomechanical and biochemical mechanisms in vivo related to the cytoskeletal conformational changes. Biomimetic mechanical stimulation, including compression, stretching, and fluid-induced shear stress, have, therefore, been used to investigate these mechanisms in vitro . Cyclic compression is able to alter MSC phenotype. MSCs subjected to dynamic compression or hydrostatic pressure induced an increase of chondrocyte lineage markers and enhanced ECM deposition.

Mechanical strain from applied loading, either unidirectional or biaxial, can direct specific stem cell lineages. For example, cyclic stretch has been demonstrated to commit MSCs to a myogenic phenotype and mouse ESCs to a vascular smooth muscle cell phenotype. Mechanical strain can also increase proliferation and inhibit differentiation in mouse and human ESCs, as well as promote cell alignment with respect to the direction of the strain. Fluid-induced shear has a significant impact on the fate of SCs, as well. In two-dimensional (2D) systems, it induces osteogenic differentiation by activating multiple intracellular signalling pathways.

MECHANICAL CONTROL OF STEM CELL RESPONSE

We provide a tentative interpretation of how mechanical cues can be used to control SC fate by applying the concept of force isotropy to synthesise the existing literature of this field.

Embryonic Stem Cells

As shown in various studies, a typical approach to mimic in vitro the native microenvironment of ESC consists in confining them, by the use of microwells or microarrays having different dimension and shape, in the aim of allowing them to form 3D microaggregates; using this method, it has been observed that higher cell densities induce cell self-renewal and maintenance of pluripotency.

Cells cultured in aggregates are characterized by a maximized cell-to-cell contact, which induces a highly isotropic cytoskeletal tension, resulting in a roundish nuclear morphology, and by low levels of oxygen concentration. Based on the existing literature, at lower levels of anisotropy of cytoskeletal tension, ESCs commit toward endoderm and mesoderm, specifically neuronal, cardiomyogenic, and haematopoietic. A highly anisotropic cytoskeletal tension directs ectodermal lineage.

Adult Stem Cells

The most studied SC type in mechanobiology are MSCs, due to their very high sensitivity to mechanical stimuli. By making a synthesis of the existing literature, it can be noted that a highly isotropic cytoskeletal tension, combined with a low oxygen concentration and with very low extracellular loads, synergistically promote MSCs self-renewal and maintenance of pluripotency. For example, in a study by Winer et al., human MSCs cultured on hydrogels having a stiffness comparable to bone marrow (0.25 kPa), have shown to self-renew and maintain pluripotency, if compared to cells cultured on more rigid substrates.

Coherently, the existing literature shows that the increase in anisotropy of cytoskeletal forces reduces pluripotency and enhances MSCs lineage commitment. For example, neural precursor cells can be obtained from MSCs by using nano/micropatterned surfaces, including nanofibres and gratings, and/or substrate having a very low stiffness, comparable to that of the brain

ECM. All these engineering strategies result in substrates which induce an isotropic cytoskeletal tension, resulting in a very roundish morphology of the cell nucleus.

SC commitment can be induced by applying biophysical cues in a synergistic manner; actually, the superimposition of extracellular loads to cells, having their own cytoskeletal tension state due to adhesive forces, can alter and regulate their fate. MSCs can express various phenotypes including fat, cartilage, muscle, fibrous tissue, and bone. Fibrous tissues, such as tendon, could be obtained by anisotropic traction forces exerted by cells and through the superimposition of anisotropic extracellular loading, such as mechanical stretching.

Myogenic and cardiomyogenic precursor cells have been obtained through a combination of different types of directional (anisotropic) extracellular loading, such as shear stress and cyclic strain, in 2D systems. Chondrogenic lineage can be achieved by a combination of intrinsic factors (e.g., a moderate substrate stiffness) and isotropic extracellular loads, including cyclic pressurisation and shear stress]. Regarding adipogenic lineage commitment, it is inhibited by anisotropic extracellular forces like mechanical strain, and favoured by isotropic-inducing factors such as a very low substrate stiffness, or a low cell density.

Conclusion

To gain a precise control over SC response and make successful translation of SC-based therapies to the clinics, the contribution of mechanical factors on SC response must be elucidated and quantified. In this paper, we have analysed the existing knowledge in this field, and we have introduced the concept of "force isotropy," as a first step towards the interpretation of the existing knowledge. Further research, in which biomechanical cues will be applied in a synergistic manner on truly 3D microenvironments, will likely provide a broader understanding of SC response to these cues.

Index